(quadratic)

THE SENSUAL ˄FORM

.

The Carus Mathematical Monographs

Number Twenty-six

(quadratic)
THE SENSUAL ∧FORM

John H. Conway
Princeton University

Assisted by
Francis Y.C. Fung

Published and Distributed by
THE MATHEMATICAL ASSOCIATION OF AMERICA

THE
CARUS MATHEMATICAL MONOGRAPHS

Published by
THE MATHEMATICAL ASSOCIATION OF AMERICA

———

Committee on Publications
James W. Daniel, *Chair*

The following Monographs have been published:

1. *Calculus of Variations*, by G. A. Bliss (out of print)

2. *Analytic Functions of a Complex Variable*, by D. R. Curtiss (out of print)

3. *Mathematical Statistics*, by H. L. Rietz (out of print)

4. *Projective Geometry*, by J. W. Young (out of print)

5. *A History of Mathematics in America before 1900*, by D. E. Smith and Jekuthiel Ginsburg (out of print)

6. *Fourier Series and Orthogonal Polynomials*, by Dunham Jackson (out of print)

7. *Vectors and Matrices*, by C. C. MacDuffee (out of print)

8. *Rings and Ideals*, by N. H. McCoy (out of print)

9. *The Theory of Algebraic Numbers*, second edition, by Harry Pollard and Harold G. Diamond

10. *The Arithmetic Theory of Quadratic Forms*, by B. W. Jones (out of print)

11. *Irrational Numbers*, by Ivan Niven

12. *Statistical Independence in Probability, Analysis and Number Theory*, by Mark Kac

13. *A Primer of Real Functions*, third edition, by Ralph P. Boas, Jr.

14. *Combinatorial Mathematics*, by Herbert J. Ryser

15. *Noncommutative Rings*, by I. N. Herstein (out of print)

16. *Dedekind Sums*, by Hans Rademacher and Emil Grosswald

17. *The Schwarz Function and its Applications*, by Philip J. Davis

18. *Celestial Mechanics*, by Harry Pollard

19. *Field Theory and its Classical Problems*, by Charles Robert Hadlock

20. *The Generalized Riemann Integral*, by Robert M. McLeod

21. *From Error-Correcting Codes through Sphere Packings to Simple Groups*, by Thomas M. Thompson

22. *Random Walks and Electric Networks*, by Peter G. Doyle and J. Laurie Snell

23. *Complex Analysis: The Geometric Viewpoint*, by Steven G. Krantz

24. *Knot Theory*, by Charles Livingston

25. *Algebra and Tiling: Homomorphisms in the Service of Geometry*, by Sherman Stein and Sándor Szabó

26. *The Sensual (Quadratic) Form*, by John H. Conway assisted by Francis Y. C. Fung

MAA Service Center
P. O. Box 91112
Washington, DC 20090-1112
800-331-1MAA   FAX: 301-206-9789

Preface

This little book is based on the Earle Raymond Hedrick Lectures that I gave at the Joint Mathematics Meetings of the American Mathematical Society and the Mathematical Association of America in Orono, Maine on August 7–9, 1991. I have been interested in quadratic forms for many years, but keep on discovering new and simple ways to understand them. The "topograph " of the First Lecture makes the entire theory of binary quadratic forms so easy that we no longer need to think or prove theorems about these forms—just look! In some sense the experts already knew something like this picture—but why did they use it only in the analytic theory, rather than right from the start?

Mark Kac's famous problem "Can one hear the shape of a drum?" when applied to n-dimensional toroidal "drums" leads to the question of which properties of quadratic forms are determined by their representation numbers. What, in other words, do we know about a lattice when we are told exactly how many vectors it has of every possible length?

Since sight and hearing were now involved, I took as the theme of the lectures the idea that one should try to appreciate quadratic forms with all one's senses, and so arose the title "THE SENSUAL FORM" for my Hedrick Lectures, and also the topics for the first two of them.

I could not settle on a single topic for the third of these lectures, even when I came to give it. So in the end, I split it into two half-hour talks: one on the shape of the Voronoi cell of a lattice, and one on the Hasse-Minkowski theory. In this book, each of these has become a fully-fledged lecture.

It was quite easy to associate the exploration of the Voronoi cell with the sense of touch. I brought smell into the act by regarding the p-adic invariants as like the individual scents of the flowers in a bouquet. After this, it was natural to end the book with a postscript which gives the reader a taste of number theory.

The book should not be thought of as a serious textbook on the theory of quadratic forms—it consists rather of a number of essays on particular aspects of quadratic forms that have interested me. The textbook I should like to write would certainly discuss the Minkowski reduction theory in more detail; also the analytic theory of quadratic forms, and Gauss's group of binary forms under composition.

There are many ways in which the treatment in this book differs from the traditional one. The "topograph " of the First Lecture is new, as are the "conorms" that are used there and in the Third Lecture, the "p-excesses" and "p-signatures" of the Fourth Lecture, and also the Gauss means that are used to prove their invariance. The p-excess of the 1-dimensional form $[a]$ is a quadratic function of a for which the corresponding symmetric bilinear function is the Hilbert norm residue symbol, $(a, b)_p$.

This reminds me of the fact that some people always smile indulgently when I mention "the prime -1", and continue to use what they presume to be the grown-up name "∞". But consider:

Every nonzero rational number is uniquely a product of powers of prime numbers p.

For distinct odd primes $\left(\frac{p}{q}\right)$ and $\left(\frac{q}{p}\right)$ differ just when $p \equiv q \equiv -1$ (mod 4).

There is an invariant called the p-signature whose definition involves summing p-parts of numbers.

If there are p-adically integral root vectors of norms k and kp, then p is in the spinor kernel.

Each of these statements includes the case $p = -1$, but none of them is even meaningful when we use the silly name "∞". In the future, I shall smile indulgently back!

Neil Sloane and I have collaborated for many years, and most of the ideas in this book first appeared in some form in one of our many papers or in our book *Sphere Packings, Lattices, and Groups.* Interested readers will find that many of the topics of this book are discussed in greater detail in those places. I am indebted to Neil for more things than I care to mention!

I thank Leonard Gillman and the Mathematical Association of America for inviting me to give the Earle Raymond Hedrick Lectures for 1991. It was Donald Albers who suggested that the notes for those lectures might be published as one of the Carus Mathematical Monographs, and so ultimately brought this book into being. I am grateful for Marjorie Senechal's continual interest in the book on behalf of the Carus Monographs committee.

If you find this book readable, the credit must largely go to Giuliana Davidoff, who as editor worked carefully through it and made a large number of helpful suggestions. Her timely intervention revitalized the writing of this book, and I am very grateful for her persistent encouragement.

But most of all, I owe Francis Fung a great debt for his offer to help in preparing those notes. This turned into a very happy collaboration. Francis was always keen to suggest that more material should go into the book, and then to shoulder the burden of helping to write it in. His ideas were always useful, and he has turned the bundles I had into the book you hold.

> *J.H.Conway,*
> *Princeton, New Jersey*

Contents

Preface vii

Note to the Reader xiii

THE FIRST LECTURE
    Can You See the Values of $3x^2 + 6xy - 5y^2$? 1

    AFTERTHOUGHTS
    $\mathbf{PSL_2(Z)}$ and Farey Fractions 27

THE SECOND LECTURE
    Can You Hear the Shape of a Lattice? 35

    AFTERTHOUGHTS
    Kneser's Gluing Method: Unimodular Lattices 53

THE THIRD LECTURE
    . . . and Can You Feel Its Form? 61

    AFTERTHOUGHTS
    Feeling the Form of a Four-Dimensional Lattice 85

THE FOURTH LECTURE
    The Primary Fragrances 91

    AFTERTHOUGHTS
    More About the Invariants: The p-Adic Numbers 117

xi

POSTSCRIPT

A Taste of Number Theory 127

References 143

Index 147

Note to the Reader

The lectures are self-contained, and will be accessible to the generally informed reader who has no particular background in quadratic form theory. The minor exceptions should not interrupt the flow of ideas. The Afterthoughts to the Lectures contain discussions of related matters that occasionally presuppose greater knowledge.

The topics are arranged so that the attention required from the reader increases slowly throught the book. Thus the First and Second Lectures should require little effort, while a reader who wants to understand the fine details of the Fourth Lecture should be prepared to do some work.

Since so much of the treatment is new to this book, it may not be easy to circumvent one's difficulties by reference to standard texts. I hope the work pays off, and that even the experts in quadratic forms will find some new enlightenment here.

Can You See the Values of $3x^2 + 6xy - 5y^2$?

Introduction

This question will lead us into the theory of quadratic forms. This is an old subject, and indeed A.-M. Legendre published most of the theory of binary quadratic forms in his *Essai sur la théorie des nombres* (1798) [Leg], while Carl Friedrich Gauss, in his monumental *Disquisitiones Arithmeticae* [Gau1] of 1801, essentially completed that theory. In this lecture, we shall present a very visual new method to display the values of any binary quadratic form. This will lead to a simple and elegant method of classifying all integral binary quadratic forms, and answering some basic questions about them.

What is a quadratic form?

Before we begin, we'd better tell you what a quadratic form is. In general, a *quadratic form* is simply a homogeneous polynomial of degree 2 in several variables, that is to say, an expression like

$$3x^2 + 6xy + y^2 - 5yz + z^2$$

in which every term has degree 2 (so there are no linear or constant terms). In this lecture, we're only going to deal with integral quadratic forms in two variables (binary forms), such as $f(x, y) = 3x^2 + 6xy - 5y^2$ where x and y and the coefficients are integers.

1

Now given $f(x,y) = 3x^2 + 6xy - 5y^2$, we can see that $f(1,1) = 3 + 6 - 5 = 4$ is a value of f. But suppose we want to know whether we can find (x,y) with $f(x,y) = 7$. This is not so easy to see immediately, and clearly it is not possible to test *all* ordered pairs (x,y).

We can get more information by changing our viewpoint a bit. We defined f to be a function on ordered pairs (x,y), but we could just as well look at it as a function of a two-dimensional vector $\mathbf{v} = x\mathbf{e}_1 + y\mathbf{e}_2$, where $\mathbf{e}_1$ and $\mathbf{e}_2$ are two linearly independent vectors. The integral linear combinations $x\mathbf{e}_1 + y\mathbf{e}_2$ form a plane lattice, as in the figure; we can add or subtract two vectors to get another.

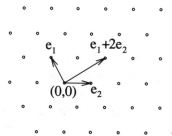

A base for the lattice is a pair such as $\mathbf{e}_1$, $\mathbf{e}_2$ (in higher dimensions, a sequence) of lattice vectors such that every lattice vector is uniquely an integral linear combination of them. Of course, the same lattice may be generated by many different bases; take $\mathbf{e}_1 + \mathbf{e}_2$ and $\mathbf{e}_2$, for example. So a binary quadratic form is a certain kind of function on a plane lattice.

In fact, a function f is a quadratic form if and only if
1) scalars behave quadratically: $f(a\mathbf{v}) = a^2 f(\mathbf{v})$, and
2) the function $B(\mathbf{v}, \mathbf{w}) := f(\mathbf{v} + \mathbf{w}) - f(\mathbf{v}) - f(\mathbf{w})$ is a symmetric bilinear form; that is,

$$B(\mathbf{w}, \mathbf{v}) = B(\mathbf{v}, \mathbf{w}),$$

and

$$B(\lambda\mathbf{u} + \mu\mathbf{v}, \mathbf{w}) = \lambda B(\mathbf{u}, \mathbf{w}) + \mu B(\mathbf{v}, \mathbf{w}).$$

In the 2×2 case, a binary form $ax^2 + hxy + by^2$ can be written more suggestively as

$$ax^2 + \frac{h}{2}xy + \frac{h}{2}yx + by^2 = \begin{pmatrix} x & y \end{pmatrix} \begin{pmatrix} a & \frac{h}{2} \\ \frac{h}{2} & b \end{pmatrix} \begin{pmatrix} x \\ y \end{pmatrix}.$$

Then the corresponding bilinear form is

$$2ax_1x_2 + hx_1y_2 + hy_1x_2 + 2by_1y_2 = 2 \begin{pmatrix} x_1 & y_1 \end{pmatrix} \begin{pmatrix} a & \frac{h}{2} \\ \frac{h}{2} & b \end{pmatrix} \begin{pmatrix} x_2 \\ y_2 \end{pmatrix}.$$

It is important to realize that the lattices in this discussion are not yet very geometrical. All that matters is their additive structure. (The expert should think of them as two-dimensional **Z**-modules.)

The two kinds of integrality

The relation between the coefficients in the quadratic form $ax^2 + hxy + by^2$ and the entries in its matrix

$$\begin{pmatrix} a & \frac{h}{2} \\ \frac{h}{2} & b \end{pmatrix}$$

leads to the two distinct definitions of what it means to be an integral quadratic form that have bedeviled this subject since its infancy. We shall say that a quadratic form is *integer-valued* just if its values for integral values of the variables are integers—this happens for $ax^2 + hxy + by^2$ exactly when a, h, b are integers. We call a quadratic form *matrix-integral* if its matrix entries are integers—$ax^2 + hxy + by^2$ is matrix integral just when a, $h/2$, b are all integers.

In number-theoretical investigations, one is most often concerned with the numbers represented by a form, and so integer-valued forms are particularly important. In geometric or algebraic contexts, the integrality of the inner products may also be important, and so there one usually needs matrix integrality.

The determinant of a lattice is the determinant of the matrix whose columns are the coordinates of a set of basis vectors. Since the determinant of the matrix of an integer-valued form is not always an integer, some authors define a modification of it called the *discriminant*. (See [Wat] for example). However, when one compares forms of varying dimensions, the discriminant is rather confusing, so we shall avoid its use. For example, we shall later meet a well-known sequence of lattices $A_0, A_1, A_2 \ldots$, whose determinants are

$$1, 2, 3, 4, 5, 6, 7, 8, 9, 10, 11 \ldots .$$

However, their discriminants as usually defined are

$$1, 1, -3, -2, 5, 3, -7, -4, 9, 5, -11, \ldots !$$

Equivalence

Two quadratic forms that look quite different may actually be essentially identical. We call two forms *(integrally) equivalent* just when they represent the same function on a lattice, but with respect to two bases. More precisely, the forms $ax^2 + hxy + by^2$ and $AX^2 + HXY + BY^2$ will be integrally equivalent just if the first is $f(x\mathbf{e}_1 + y\mathbf{e}_2)$ and the second is $f(X\mathbf{f}_1 + Y\mathbf{f}_2)$, where $\{\mathbf{e}_1, \mathbf{e}_2\}$ and $\{\mathbf{f}_1, \mathbf{f}_2\}$ are two bases for the same lattice.

This entails in particular that the forms take on exactly the same values as the variables range over the integers. For example, if

$$\mathbf{e}_1 = \mathbf{f}_2, \qquad \mathbf{e}_2 = -\mathbf{f}_1 - \mathbf{f}_2,$$

then

$$x\mathbf{e}_1 + y\mathbf{e}_2 = -y\mathbf{f}_1 + (x - y)\mathbf{f}_2 = w\mathbf{f}_1 + z\mathbf{f}_2,$$

so

$$2x^2 - 4xy + 3y^2 = 2(x - y)^2 + y^2 = 2z^2 + w^2,$$

where $z = x - y$ and $w = -y$. Now as x and y range independently over the integers, z and w do likewise; we see that the values of the form $2x^2 - 4xy + 3y^2$ and those of the form $2z^2 + w^2$ are the same.

In matrix terms, we have

$$\begin{pmatrix} w \\ z \end{pmatrix} = \begin{pmatrix} 0 & -1 \\ 1 & -1 \end{pmatrix} \begin{pmatrix} x \\ y \end{pmatrix}$$

so

$$(w \quad z) = (x \quad y) \begin{pmatrix} 0 & 1 \\ -1 & -1 \end{pmatrix},$$

and

$$\begin{aligned}
2z^2 + w^2 &= (w \quad z) \begin{pmatrix} 1 & 0 \\ 0 & 2 \end{pmatrix} \begin{pmatrix} w \\ z \end{pmatrix} \\
&= (x \quad y) \begin{pmatrix} 2 & -2 \\ -2 & 3 \end{pmatrix} \begin{pmatrix} x \\ y \end{pmatrix} \\
&= 2x^2 - 4xy + 3y^2.
\end{aligned}$$

More generally, two matrices A and B represent the same form with respect to different bases just when there exists a matrix M such that $M^T A M = B$ and both M and M^{-1} have integer entries. (Or, equivalently, M has integral entries, and $\det(M) = \pm 1$.) In particular, we see that the determinant of a form is independent of choice of base: $\det(B) = \det(A)(\det(M))^2 = \det(A)$.

Primitive vectors, bases and superbases, strict and lax

To understand a quadratic form, we should study it as far as possible in a manner independent of base. Our theory reduces to the detailed study of three concepts: primitive vectors, bases, and superbases. Since $f(k\mathbf{v}) = k^2 f(\mathbf{v})$, to explore the values of f at all vectors it will suffice to explore its values at *primitive vectors*, those not of the form $k\mathbf{v}$ for any integer $k > 1$. Also, since $f(-\mathbf{v}) = f(\mathbf{v})$ it will often be convenient to think in a loose way of $\mathbf{v}$ and $-\mathbf{v}$ as the "same" vector. A *strict vector* will be a primitive vector $\mathbf{v}$ counted as distinct from $-\mathbf{v}$, and a *lax vector* is a pair $\pm\mathbf{v}$ where $\mathbf{v}$ is a strict vector.

In the same spirit, a *strict base* is an ordered pair $(\mathbf{e}_1, \mathbf{e}_2)$ whose integral linear combinations are exactly all the lattice vectors. A *lax base* is a set $\{\pm\mathbf{e}_1, \pm\mathbf{e}_2\}$ obtained from a strict base.

Finally, a *strict superbase* is an ordered triple $(\mathbf{e}_1, \mathbf{e}_2, \mathbf{e}_3)$, for which $\mathbf{e}_1 + \mathbf{e}_2 + \mathbf{e}_3 = 0$ and $(\mathbf{e}_1, \mathbf{e}_2)$ is a strict base(i.e., with strict vectors), and a *lax superbase* is a set $\{\pm\mathbf{e}_1, \pm\mathbf{e}_2, \pm\mathbf{e}_3\}$ where $(\mathbf{e}_1, \mathbf{e}_2, \mathbf{e}_3)$ is a strict superbase. From now on we'll usually use the "lax" notions without further comment. We remark that all the vectors that appear in bases and superbases are primitive; conversely, every primitive vector does appear in some base.

The relation between bases and superbases

Plainly each superbase

$$\{\pm e_1, \pm e_2, \pm e_3\}$$

contains just three bases

$$\{\pm e_1, \pm e_2\}, \{\pm e_1, \pm e_3\}, \{\pm e_2, \pm e_3\}.$$

On the other hand, each base $\{\pm e_1, \pm e_2\}$ is in just two superbases,

$$\{\pm e_1, \pm e_2, \pm(e_1 + e_2)\}, \{\pm e_1, \pm e_2, \pm(e_1 - e_2)\}.$$

Note that each of these really *is* a superbase.

The topography of bases and superbases

We can make a picture to describe the incidences between bases and superbases. We draw a graph joining each superbase (○) to the three bases (□) in it.

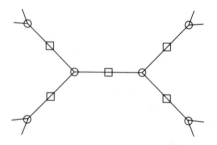

Since each base is in just two superbases, this *topograph* picture can be regarded as having an edge for each base and a vertex for each superbase.

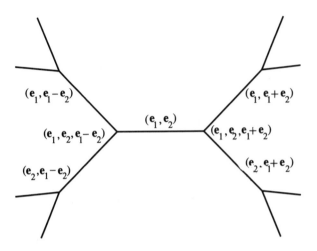

Where are the primitive vectors in the topograph?

If $\pm e_1$ is in a base $\{\pm e_1, \pm e_2\}$ which is in turn in a superbase $\{\pm e_1, \pm e_2, \pm e_3\}$, then $\pm e_1$ is in just one of the other two bases in the superbase, namely $\{\pm e_1, \pm e_3\}$. So in our picture, in which we have suppressed the $\pm$'s, the nodes and edges that involve e_1 form a path. We can therefore add a face bounded by this path to our topo-

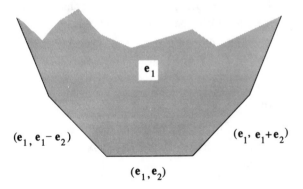

graph and identify it with $\pm e_1$ (so that the picture becomes more like a travel map on a surface). In the resulting fully labelled topograph, each region is labelled with a (lax) vector $\pm v$ (but we usually omit the sign), two vectors separated by an edge form a (lax) base, and three around a vertex form a (lax) superbase.

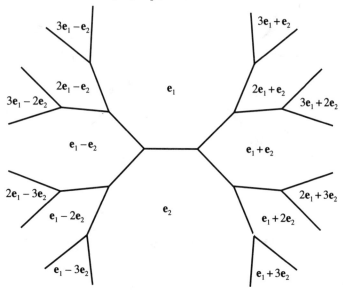

Norms of vectors

Up to this point in our discussion of the topograph, the values of f haven't even been mentioned. (So we see that the shape of the topograph does not depend on f.) We now fix on a particular quadratic form f and, for this f, call $f(\mathbf{v})$ the *norm* of $\mathbf{v}$.

The arithmetic progression rule

Suppose we know the values of a quadratic form f at the three vectors $\{\mathbf{e}_1, \mathbf{e}_2, \mathbf{e}_3\}$ of some superbase. How do we find its values elsewhere? We use the formula

$$f(\mathbf{v}_1 + \mathbf{v}_2) + f(\mathbf{v}_1 - \mathbf{v}_2) = 2[f(\mathbf{v}_1) + f(\mathbf{v}_2)],$$

which is essentially equivalent to a well-known geometrical theorem of Apollonius. To verify this, let $B(\mathbf{v}_1, \mathbf{v}_2)$ be the bilinear form associated to f. Then

$$f(\mathbf{v}_1 + \mathbf{v}_2) + f(\mathbf{v}_1 - \mathbf{v}_2)$$
$$= f(\mathbf{v}_1) + f(\mathbf{v}_2) + B(\mathbf{v}_1, \mathbf{v}_2) + f(\mathbf{v}_1) + f(\mathbf{v}_2) - B(\mathbf{v}_1, \mathbf{v}_2)$$
$$= 2\big[f(\mathbf{v}_1) + f(\mathbf{v}_2)\big].$$

This formula tells us that if

$$a = f(\mathbf{v}_1), \quad b = f(\mathbf{v}_2), \quad c = f(\mathbf{v}_1 + \mathbf{v}_2), \quad d = f(\mathbf{v}_1 - \mathbf{v}_2)$$

are the values of f in the four regions around an edge of the topograph, then d, $a + b$, c is an arithmetic progression (with a common difference that we'll call h). For this reason, we call this the *Arithmetic Progression Rule*. Also, we mark each directed edge with the appropriate h (the difference of the arithmetic progression). We can choose the direction so as to avoid a negative value for h, and we usually omit the arrow when h is 0. In summary:

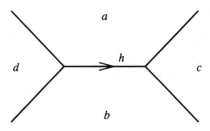

means that $c = (a + b) + h$, $d = (a + b) - h$ (see [Sel], [Vor], [Hur]), while

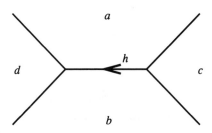

means that $c = (a + b) - h$, $d = (a + b) + h$.

Writing quadratic forms with respect to a base

Now say we're given the values of a quadratic form f at a superbase:

$$f(\mathbf{e}_1) = a, \quad f(\mathbf{e}_2) = b, \quad f(\mathbf{e}_1 + \mathbf{e}_2) = c.$$

Then we write

$$\mathbf{v} = x\mathbf{e}_1 + y\mathbf{e}_2$$

so that

$$f(\mathbf{v}) = f(x, y)$$
$$= ax^2 + (c - a - b)xy + by^2,$$

since this function is the only homogeneous quadratic that has

$$f(1, 0) = a, \quad f(0, 1) = b, \quad \text{and} \quad f(1, 1) = c.$$

Notice that the coefficient of xy is the difference term h that appears in the Arithmetic Progression Rule.

So given a superbase and any three integers a, b, and c, we can find an integral quadratic form with values a b, and c at that superbase.

Now the matrix representation of f with respect to the base corresponding to the horizontal edge in the previous figure is:

$$(\, x \quad y \,) \begin{pmatrix} a & \frac{1}{2}h \\ \frac{1}{2}h & b \end{pmatrix} \begin{pmatrix} x \\ y \end{pmatrix}$$

and so the determinant of f is $ab - (\frac{1}{2}h)^2$. Recall that since an invertible change of base has determinant ± 1, the determinant is an invariant of integral equivalence.

The tree property

In fact, our topograph turns out to be a *tree*; that is, a connected graph with no circuits. We shall prove this in the next two sections.

Why is there no circuit? Construct the form that takes on the values 1, 1, 1 at the superbase corresponding to a vertex P.

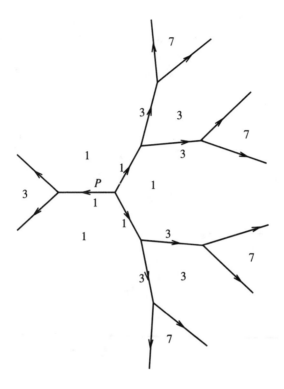

It seems that the numbers increase as we walk away from P. Does this continue?

The Climbing Lemma.   *Suppose a, b, and h in the figure below are positive. Then the third number c at Q is also positive, and the edges that emerge from Q both point away from Q.*

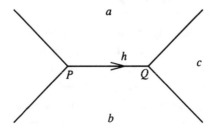

Proof. The number c equals $a + b + h$ by the Arithmetic Progression Rule. Then one of the edges going away from Q is between faces marked a and $a + b + h$, which sum to $b + (2a + h)$. Since it came away from a face marked b, the appropriate edge label is $2a + h$.

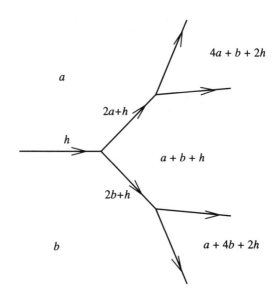

Similarly, the other edge is marked $2b + h$. So if we have an arrow going positively from P to Q between positive numbers, then the arrows going out of Q have the same property, but with larger numbers all around. So if we continue walking away from P in this direction, we can never return to P because all of the numbers involved keep getting bigger and bigger.   □

Now here's the important point: the shape of the topograph does not depend on the quadratic form, since it only depends on bases and superbases for the underlying lattice $\mathbf{Z}^2$. So the proof shows that there is no cycle in the picture for *any* quadratic form, even though we proved it by referring to a particular quadratic form. Is the topograph connected? As far as we're concerned, we've drawn the part of the picture that we can get by continuing from one particular node; there might be some other superbases that we just can't get to. We shall see that this is not so in the next section by proving a lemma about positive definite forms.

Wells for positive definite forms

The form f is called *positive semidefinite* if $f(\mathbf{v}) \geq 0$ for all $\mathbf{v}$ and *positive definite* if $f(\mathbf{v}) > 0$ for all $\mathbf{v} \neq \mathbf{0}$. *Negative semidefinite* and *negative definite* are similarly defined.

Let's look at the topography of the (positive definite) quadratic form whose values at some superbase are 5, 27, 55. The Climbing Lemma shows that if we walk away from P through either Q or R, the numbers will increase. So instead we step down to S, at which the values are 5, 9, 27. Repeating this process, we find ourselves stepping down against the flow along the dashed path $STUW$ in the figure.

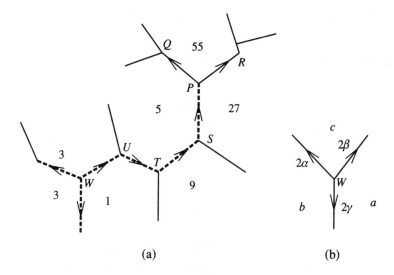

(a)   (b)

We stop at W because all three arrows at W point away from W. We call a superbase W as in the figure (b) a *well* if the edge-marks 2α, 2β, and 2γ are all associated to arrows pointing away from W. If a, b, c are the values of f at this superbase, then from the Arithmetic

Progression Rule we have

$$2\alpha = b + c - a, \quad 2\beta = c + a - b, \quad 2\gamma = a + b - c,$$

and so

$$a = \beta + \gamma, \quad b = \alpha + \gamma, \quad c = \beta + \alpha.$$

The process above shows that there exists a well for any positive definite integral quadratic form.

The Well Lemma.   Suppose we have a well for a positive definite quadratic form or, equivalently, a superbase surrounded by three positive numbers that satisfy the "triangle inequality"; i.e., $a + b \geq c$, $a + c \geq b$, and $b + c \geq a$. Then the three vectors in this superbase are in fact the three primitive vectors of smallest norm (that is, where f takes its smallest value).

Proof.   First, let α, β, and γ be the three nonnegative numbers

$$\alpha = \frac{b + c - a}{2}, \quad \beta = \frac{c + a - b}{2}, \quad \gamma = \frac{a + b - c}{2}.$$

We write the general vector $\mathbf{v}$ as

$$\mathbf{v} = m_1 \mathbf{e}_1 + m_2 \mathbf{e}_2 + m_3 \mathbf{e}_3$$

and first ask what is the formula for f in terms of the m_i. Since $\mathbf{e}_1 + \mathbf{e}_2 + \mathbf{e}_3 = \mathbf{0}$, $\mathbf{v}$ also equals

$$(m_1 - k)\mathbf{e}_1 + (m_2 - k)\mathbf{e}_2 + (m_3 - k)\mathbf{e}_3,$$

and so f can only depend on the differences of the m_i.

Now the linear combinations of

$$(m_2 - m_3)^2, \quad (m_1 - m_3)^2, \quad \text{and} \quad (m_1 - m_2)^2$$

constitute a three-dimensional space of quadratic forms. Hence if any two of the m_i are equal, then $\mathbf{v}$ is a multiple of the remaining basis vector. So we shall suppose that all of the m_i are distinct.

Now, we can verify *Selling's formula*:

$$f(\mathbf{v}) = \alpha(m_2 - m_3)^2 + \beta(m_1 - m_3)^2 + \gamma(m_1 - m_2)^2.$$

(This is valid because the right-hand side is a quadratic form that agrees with f at the superbase e_1, e_2, e_3; alternatively, just write the matrix of the quadratic form with respect to the given base and match terms). If v is not a multiple of one of the e_i, all of the differences $m_i - m_j$ are nonzero, so $f(v) \geq \alpha + \beta + \gamma$ which is at least as big as each of $a = \beta + \gamma$, $b = \alpha + \gamma$, $c = \alpha + \beta$. Indeed, it is strictly bigger unless one of α, β, γ is 0. So in any case, the values at the well are the three smallest primitive values of the form, and if α, β, γ are nonzero, then its value anywhere else is strictly larger. □

Selling's formula[Sel] was generalized to arbitrary dimensions by Voronoi [Vor]: if $e_0, \ldots, e_n$ is a superbase, with $e_i \cdot e_j = p_{ij}$, then $f(\sum m_i e_i) = \sum_{i<j} p_{ij}(m_i - m_j)^2$. (See also [CSVI] for more information).

The topograph is connected

To see that the topograph is connected, take the particular positive definite quadratic form that has a well with $\alpha = \beta = \gamma = 1$ at some chosen superbase. Then, by climbing down, we see that any component of the topograph must contain a well, the three vectors of which must be those that yield the three smallest primitive values of the form. This well can only be "our" well. Therefore the component must be "our" component, and so the topograph is indeed connected.

Conorms, vonorms, simple and double wells

The three numbers a, b, and c, we call the *vonorms* of f ("Voronoi norms"—they are the norms of the Voronoi vectors of f, which we will define formally in Lecture Three). The three numbers α, β, and γ are the *conorms* of f (they have also been called the Selling parameters of f, but in higher dimensions, these concepts do not agree). Either set of numbers determines the other:

$$a = \beta + \gamma \quad b = \alpha + \gamma \quad c = \alpha + \beta$$

and

$$\alpha = \frac{b+c-a}{2}, \quad \beta = \frac{a+c-b}{2}, \quad \gamma = \frac{a+b-c}{2}.$$

We have seen that when α, β, $\gamma > 0$, there is a unique well. In the picture, the marks on all the edges are strictly positive and all the arrows point away from the well. We call this a *simple well*.

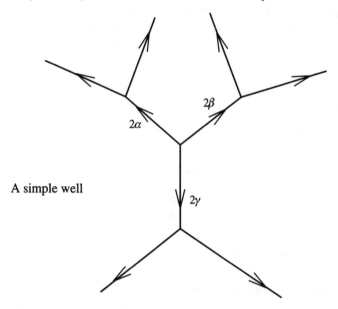

A simple well

If a well is not simple, then without loss of generality $\alpha = b$, $\beta = a$, $\gamma = 0$, and Selling's formula reads

$$f(\mathbf{v}) = b(m_2 - m_3)^2 + a(m_1 - m_3)^2.$$

So the value of f at $m_1\mathbf{e}_1 + m_2\mathbf{e}_2$ is $am_1^2 + bm_2^2$, and at the four lax vectors

$$\pm\mathbf{e}_1, \quad \pm\mathbf{e}_2, \quad \pm(\mathbf{e}_1 + \mathbf{e}_2), \quad \pm(\mathbf{e}_1 - \mathbf{e}_2)$$

the values are

$$a, \qquad b, \qquad a+b, \qquad a+b$$

and everywhere else its values are strictly larger.

In the picture below, the edge between a and b has a well at each end, and no arrow (or an arrow labeled 0). Every other edge has an arrow pointing away from this edge. We call this a *double well*.

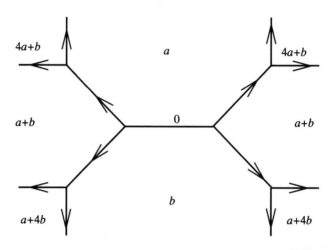

A double well

Our next two figures illustrate particular forms having a simple and a double well.

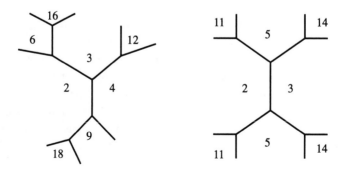

Classifying forms by signs

It turns out to be a good idea to separate binary integer-valued quadratic forms according to the signs of the numbers that they represent at nonzero primitive vectors:

The + Forms: These are the positive definite forms we have just discussed—their smallest values surround a simple or double well, and increase as we go away.

The − Forms: The negative definite forms are of course very similar to the positive definite ones.

The +− Forms: These are an important family. We'll discuss their topography in the next section. We shall see that there is a periodic "river" that separates the positive and negative values of the form.

The 0 Form: When the only value is 0, the topograph consists of infinitely many "lakes" labeled 0.

The 0+ Forms: A form that takes on only 0 and positive values is equivalent to a scalar multiple of the form x^2. We'll see that the values increase as we move away from a single lake (value 0) that is surrounded by regions all of the same value.

The 0− Forms are similar to the 0+ forms.

The 0+− Forms are the last case. Here we shall show that there are two distinct "lakes" of value 0, joined by a finite river that separates the positive and negative values.

Indefinite forms not representing 0: The river

The form $\begin{pmatrix} a & h \\ h & b \end{pmatrix}$ is of this kind just if its determinant $ab - h^2$ is negative, but not the negative of a perfect square. For such a form, the topograph necessarily contains an edge lying directly between a positive and negative value. (This is because it is connected, so that we can walk from a place where the form is positive to a place where the form is negative, and 0 is not represented). In practice, if we are given a superbase S at which the values are positive, we can reach such an edge by climbing down, as in the figure.

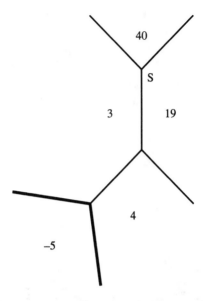

In our pictures for such forms we shall thicken lines that separate positive numbers from negative ones, and call them *river edges*. Now if at a superbase P we have a negative value a and a positive value b (and hence a river edge), then the third value c must be either positive or negative, and so there is a second river edge at P, say PQ.

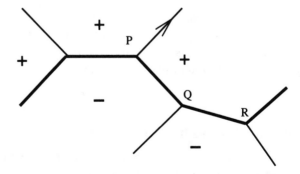

We see therefore that each river edge meets another one at each of its ends. In this way we get a path $PQR\ldots$ that separates the positive and negative regions; we call this the *river*.

The Climbing Lemma shows that if we climb away from the river on the positive side, the values will continually increase. (This is because an edge, such as PS in the diagram below, leading directly away from the river on this side receives a label of $h = c + a - b$ in which $a > 0$, $b < 0$, $c > 0$.) Similarly, if we move away from the river on the other side, the values get more and more negative.

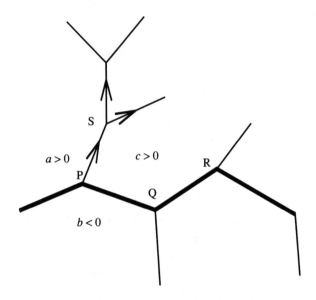

In other words: if you stray from the river, the values go up (in absolute value). Notice that this proves that the river is unique, because the topograph is connected and if you move away from our river, you'll see values of only one sign. So you won't get to another river.

Integer-valued forms have periodic rivers

For the form $x^2 + 6xy - 3y^2$, after moving along the river from the initial superbase P_0 surrounded by the numbers 1, 2, -3, we find another superbase P_1 surrounded in exactly the same way. If we move the same distance again, we shall see yet another such superbase P_2 and so on; the surroundings of the river repeat periodically. We shall

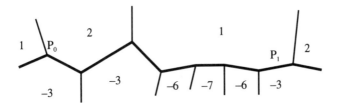

show that if the coefficients a, h, b, in the $+-$ form $ax^2 + hxy + by^2$ are integers, then the river is necessarily periodic in this way. The

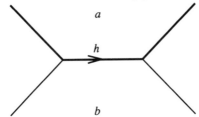

determinant of this form is $d = ab - (\frac{1}{2}h)^2$. If we suppose that the base corresponds to an edge of the river, then exactly one of a and b is negative, so ab is negative, and $|d| = (\frac{1}{2}h)^2 + |ab|$. Hence $|\frac{1}{2}h| < \sqrt{|d|}$, and $|ab| = |d| - (\frac{1}{2}h)^2$, so there are only finitely many values for a, b, and h. Now a, b, and h together determine all the other labels, and there are only finitely many possible triples (a, b, h) for river edges. So some two edges of the river must be surrounded in the same way, and the river must be periodic (since the values of a, b, and h at any edge determine the entire topography).

Now, *can* we really see the values of

$$f(x, y) = 3x^2 + 6xy - 5y^2?$$

For this form, we have $f(1, 0) = 3$, $f(0, 1) = -5$, and $f(1, 1) = 4$. So we start from a superbase labeled $(4, 3, -5)$ and sail up the river.

On the left bank of this particular river, we see that the negative primitive values

$$-5, -8, -24, -29, -53, -60, -69, -92, -101, \ldots$$

arise from essentially just two shapes of tree (the upper tree also arises in a mirror-image version). Its positive primitive values

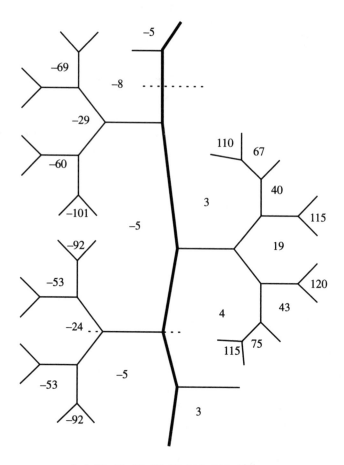

$$3, 4, 19, 40, 43, 67, 75, 110, 115, 120, \ldots$$

nestle between the branches of essentially just one shape of tree (and its mirror image). If we multiply these numbers by squares, we obtain all the values of the form. We see in particular that the Diophantine equations

$$3x^2 + 6xy - 5y^2 = 7 \quad \text{and} \quad 3x^2 + 6xy - 5y^2 = -100$$

are insoluble in integers.

We can also see in this figure the isometry group of f (which by definition consists of those linear transformations that preserve f).

This is an infinite dihedral group generated by two reflections in the dotted lines of the figure; one bisecting an edge between the regions of values -8 and 3, the other running between two regions of value -5.

Semidefinite forms

A *lake* is the region corresponding to a vector where the form represents 0. Then the Arithmetic Progression Rule tells us that the values in the regions around a lake form an infinite arithmetic progression, as in figure (a).

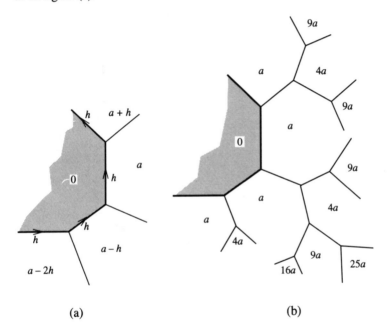

(a)   (b)

For the 0 form, all the values are 0, and the topograph consists just of infinitely many lakes. Now a $0+$ form represents 0 and positive numbers only. For any such form, the arithmetic progression around the lake must be constant, or else it would contain a negative number. So in fact the form is a scalar multiple of x^2, as we see in figure (b) above. A similar discussion holds for $0-$ forms.

Indefinite forms representing 0

Finally we come to the interesting case of $0+-$ forms, which represent numbers of all three signs.

Now we have a lake, and a nonconstant arithmetic progression, which must change sign somewhere around the lake shore. If the change is directly between positive and negative, it happens at an edge of some river flowing out from the lake. For an integer-valued form, this river must end by flowing into another lake, since if it were infinite, it would be periodic by our previous argument.

In the example below, we start from a superbase labeled $(0, 6, -5)$, and work towards one labeled $(2, -9, 0)$.

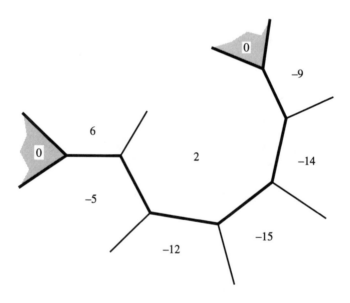

There is a special case in which the river is of zero length. This happens when the arithmetic progression contains 0. The form is then equivalent to axy, and the topograph has two lakes abutting along an edge—the "weir"—with positive values at one end and negative ones at the other.

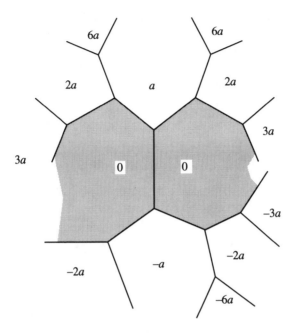

A form with a weir

Valediction

In a very real sense, the "topograph" described in this lecture really does convey the entire set of values of the quadratic form in a vivid and visual way. This is useful not merely for orientation: theorems that once had to be proved algebraically or arithmetically can now become so obvious that they no longer require proof. Perhaps the most important one summarizes the content of this lecture:

Theorem. For any given integers a, b, h, n, there is an algorithmic way to decide whether the Diophantine equation

$$ax^2 + hxy + by^2 = n$$

is solvable in integers (x, y), and to find such integers in the case when it is solvable. There is also an effective way to solve the equivalence problem for such forms and to find their isometry groups.

This completes our classification of integral binary quadratic forms. In the next lecture, we'll concentrate on higher-dimensional lattices, which are linked by their length functions to positive definite forms.

Later we will return to vonorms and conorms, with their mysterious relationships like $\gamma = \frac{a+b-c}{2}$.

$\mathbf{PSL_2(Z)}$ and Farey Fractions

Introduction

The afterthoughts following our lectures will add more detail, introduce some related topics, or just put our ideas into some other context. We shall occasionally presume some knowledge of more standard treatments. The underlying subject of this lecture is the group $\mathbf{PSL_2(Z)}$, which can be regarded as the set of all maps

$$z \mapsto \frac{az + b}{cz + d}, \qquad a, b, c, d \in \mathbf{Z}, ad - bc = 1.$$

from the upper half-plane to itself. It is interesting to see how our topograph is drawn in the upper half plane $H = \{x + iy | y > 0\}$.

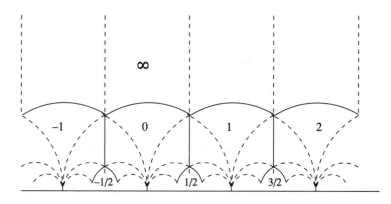

The picture shows H divided into fundamental regions for the group $\mathbf{PSL}_2(\mathbf{Z}) = \Gamma$. The solid edges form a tree with three edges per vertex whose nodes and edges correspond to the superbases and bases for $\mathbf{Z}^+$. Each of the regions of our topograph consists of "fans" of fundamental regions.

We draw one such fan by itself:

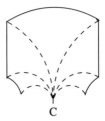

C

The fan labeled p/q is the face corresponding to the primitive vector (p, q) in $\mathbf{Z}^2$. It happens that the center C of this fan is the rational number p/q. Note that since $-p/-q$ is the same rational number as p/q, the primitive vectors $(-p, -q)$ and (p, q) automatically correspond to the same fan.

The geometry of this figure (which is really hyperbolic non-euclidean geometry) adds quite a bit to our knowledge. The fans have inscribed circles, which we now show:

$$\infty = 1/0$$

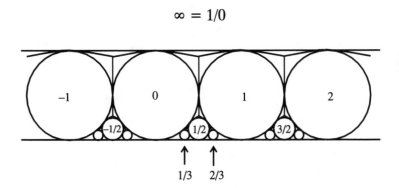

and next we draw these circles by themselves:

$$\infty = 1/0$$

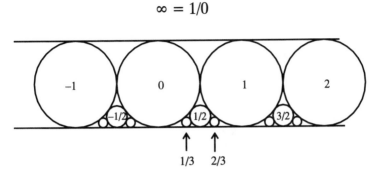

They are usually called *Ford circles*. The Ford "circle" for $\infty = 1/0$ is the horizontal line at height 1. The Ford circle for p/q is the circle of diameter $1/q^2$ in H that touches the real axis at p/q. The Farey series of order d consists of every rational number whose denominator is at most d. These correspond to the Ford circles that intersect any horizontal line L at height between $1/d^2$ and $1/(d + 1)^2$. So for example from the line below, we get the Farey series of order 4, namely $\ldots \frac{0}{1}, \frac{1}{4}, \frac{1}{3}, \frac{1}{2}, \frac{2}{3}, \frac{3}{4}, \frac{1}{1} \cdots$.

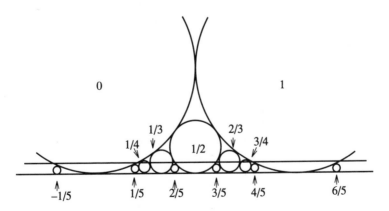

The "mediant" rule for Farey series tells us the first fraction that will appear between the adjacent Farey fractions p/q and r/s as the order is suitably increased; this is the *mediant fraction* $(p+r)/(q+s)$.

This mediant operation is more easily understood when one realizes that the fraction p/q really represents the vector (p, q). The situation is pictured below in the topograph.

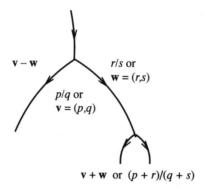

This topograph shows that if the regions on either side of an edge have labels p/q and r/s, then those at the ends of that edge will have labels $(p \pm r)/(q \pm s)$. So the first fraction between p/q and r/s with larger denominator than those will indeed be $(p + r)/(q + s)$, their mediant.

Some theorems of Diophantine approximation also become obvious. For example, for any irrational real number α, there are infinitely many rational numbers p/q for which $|\alpha - \frac{p}{q}| \le 1/2q^2$.

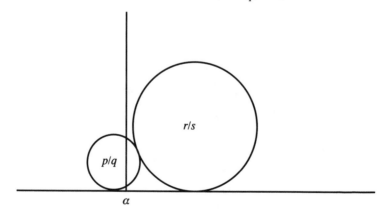

This is because if we take any two adjacent circles p/q and r/s whose tangent points with **R** lie on opposite sides of α then the vertical line through α must hit at least one of them. But if, as drawn in the figure, it hits the Ford circle for p/q, then $|\alpha - \frac{p}{q}| \leq 1/2q^2$.

The action of the groups $\mathbf{SL_2(Z)}$ and $\mathbf{GL_2(Z)}$ on the rationals becomes much easier to visualize if we apply the conformal map $z \mapsto \frac{z-i}{z+i}$ to change from the upper half-plane to the Poincaré disc, as in our next three figures. Now the Ford circles become circles tangent to the bounding disc. The group $\mathbf{GL_2(Z)}$ consists of all the symmetries of this figure, while $\mathbf{SL_2(Z)}$ consists only of the "rotational" ones.

The first of these figures exhibits the order 4 subgroup of $\mathbf{GL_2(Z)}$ generated by the operations $t \mapsto -t$ and $t \mapsto 1/t$, while the second one exhibits the order 6 subgroup generated by $t \mapsto 1/t$ and $t \mapsto 1-t$. In fact, $\mathbf{GL_2(Z)}$ is the free product of these two finite groups, amalgamated over their common subgroup of order 2. (If we pass to the rotation subgroups, we see how $\mathbf{SL_2(Z)}$ arises as the free product of groups of orders 2 and 3.)

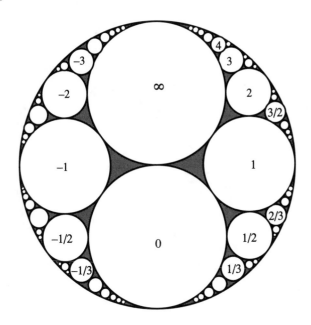

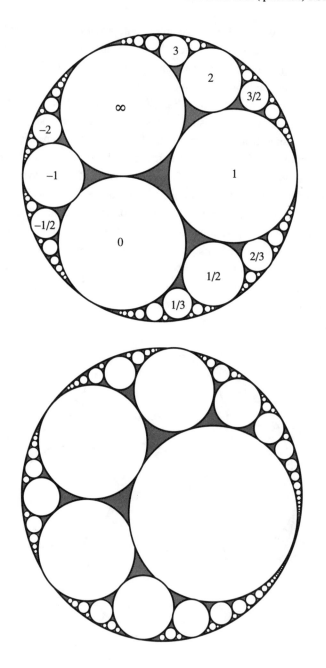

These two figures display some finite subgroups of **PGL**$_2$(**Z**) by Euclidean congruences. However, all the symmetries are represented by congruences in non-Euclidean geometry. In fact, **PGL**$_2$(**Z**) is the full symmetry group of a packing of the hyperbolic plane by circles of infinite radius (called *horocycles*). There is one horocyle for each rational number p/q, and it touches the boundary (identified with the real axis) at p/q. We have already seen this from three special viewpoints which exhibit particular subgroups; our last figure shows it from a viewpoint that has no special symmetry.

Can You Hear the Shape of a Lattice?

Introduction

Our title is intended to recall Mark Kac's famous question "Can one hear the shape of a drum?" Kac's article [Kac] drew wide attention to an important old problem which was first raised about 100 years ago. In physical language, we may state this as "do the frequencies of the normal modes of vibration determine the shape of the drum?" Of course this is a purely mathematical problem:—do the eigenvalues of the Laplacian for the Dirichlet problem determined by a planar domain determine the shape of that domain?

When the titles for the lectures on which this book is based were chosen, this problem was still unsolved. By the time they were given, it had been solved by Gordon, Webb, and Wolpert, who made use of some previous work by Sunada and Buser.

It is always exciting to see a classical problem solved, and in this case the solution can be made particularly easy, so although it has little to do with the main topic of these lectures, we give a simple solution to the Kac problem in this lecture.

It is perhaps fortunate that the solution took so long to find, because the consideration of the problem has led to a lot of interesting mathematics. In particular, one can consider the problem for arbitrary Riemannian manifolds; that is, surfaces of possibly arbitrary dimension with an appropriate metric. Very soon after Kac's lecture, John Milnor found the first counterexamples in this more general setting; namely, two 16-dimensional tori that "sound the same" (in the sense that their Laplacians have the same eigenvalues) although they are of different shapes[BGM].

Isospectral lattices

What does this have to do with lattices? A lattice L in $\mathbf{R}^n$ is the set of integral linear combinations of n linearly independent vectors $\mathbf{v}_1, \mathbf{v}_2, \ldots, \mathbf{v}_n$. Now given a lattice in some Euclidean space, there is a way to "roll up" the space around the lattice and get a quotient manifold. As an example, we take $\mathbf{Z}^2$ in the Euclidean space $\mathbf{R}^2$ and identify two points of $\mathbf{R}^2$ if their difference lies in the lattice $\mathbf{Z}^2$. So the quotient space $\mathbf{R}^2/\mathbf{Z}^2$ is a torus, as is probably familiar to the reader. In a similar way, we can construct higher-dimensional tori as the quotients of other Euclidean spaces by lattices contained in them.

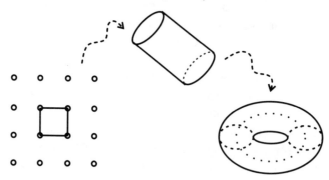

These tori are Riemannian manifolds since they "inherit" the Euclidean space metric. Now in $\mathbf{R}^{16}$, there are two dissimilar lattices, $E_8 \oplus E_8$ and D_{16}^+, whose quotient tori are *isospectral* —that is, "they sound the same". This is because the spectrum of the quotient manifold turns out to be determined entirely by the number of vectors of each length in the lattice[GHL], and so by its so-called *theta function*:

$$\Theta_L(q) = \sum_l a_l q^l, \qquad q = e^{2\pi i z},$$

where a_l is the number of vectors in L of squared length l. After this result, it is natural to say that a lattice property is *audible* if it is

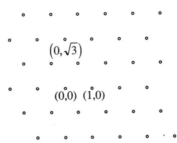

determined by the theta function. For instance, the hexagonal lattice has one vector $\mathbf{v}$ whose squared length $N(\mathbf{v})$ (or *norm*) is 0, six of norm 1, then six of norm 3, six of norm 4, twelve of norm 7, and so on, so that its theta function is

$$1 + 6q + 6q^3 + 6q^4 + 12q^7 \ldots.$$

The main problem of this chapter is to find in which dimensions there can be two dissimilar lattices that have the same theta function, or equivalently the same number of vectors of every length. This is a purely geometrical problem, although it arose in an analytical context, and Milnor's argument used some other analytical ideas, as follows.

Milnor's example

It is known that the theta functions of even unimodular lattices (those in which every squared length is even, and which have one point per unit volume) are highly restricted—they are very special functions called *modular forms* for the full modular group $\mathbf{PSL}_2(\mathbf{Z}) = \Gamma$. This means that

$$\Theta\left(\frac{az+b}{cz+d}\right) = (cz+d)^{n/2}\Theta(z)$$

for every 2×2 matrix $\begin{bmatrix} a & b \\ c & d \end{bmatrix}$ of integers with determinant 1, where n is the dimension of the lattice. In the 16-dimensional case there happens to be just one such function up to scalars: it looks like

$$\Theta(q) = 1 + 480 \sum_n \sigma_7(n)q^{2n}$$

where $\sigma_7(n)$ is the sum of the 7th powers of the divisors of n. So every 16-dimensional even unimodular lattice must have exactly this theta function (since the coefficient of q^0 is 1 in both cases). Witt showed in [Wi] that there are just two even unimodular 16-dimensional lattices, now usually called $E_8 \oplus E_8$ and D_{16}^+, and so we see that the two 16-dimensional tori found by Milnor, namely

$$\mathbf{R}^{16} \Big/ E_8 \oplus E_8 \quad \text{and} \quad \mathbf{R}^{16} \Big/ D_{16}^+$$

are isospectral. Thus, the shape of a 16-dimensional manifold need not be audible. By considering other pairs of lattices with the same theta functions, we can obtain isospectral tori in differing dimensions. In the next few sections we shall describe some of the lattices involved.

The 16-dimensional lattices

The root lattice D_n consists of all vectors $(x_1, x_2, \ldots, x_n)$ for which the x_i are integers with even sum. The n-dimensional *diamond packing* D_n^+ is the set of points obtained by adjoining to D_n its coset $D_n + (\frac{1}{2}, \frac{1}{2}, \ldots, \frac{1}{2})$. We call it the diamond packing because the points of D_3^+ correspond to the positions of carbon atoms in a diamond. D_n^+ can only be a lattice when n is even, because when n is odd, the vector $(\frac{1}{2}, \frac{1}{2}, \ldots, \frac{1}{2}) + (\frac{1}{2}, \frac{1}{2}, \ldots, \frac{1}{2}) = (1, 1, \ldots, 1)$ is not in D_n^+. For even n, D_n^+ is a lattice, and it is an even unimodular lattice when n is a multiple of 8. In general, the symmetries of D_n^+ are just the obvious ones: we can permute the coordinates in any way, and we can change the sign of any even number of coordinates. However, when $n = 8$, there are more symmetries, and so it is appropriate to use another name, E_8, for D_8^+. (This lattice is the root lattice of the Lie algebra of type E_8). The two isospectral lattices in 16 dimensions are $E_8 \oplus E_8$ and D_{16}^+.

At the end of this book, we'll prove that even unimodular lattices exist only when the dimension is a multiple of 8. In 8 dimensions we have only E_8, and in 16 dimensions the above lattices $E_8 \oplus E_8$ and D_{16}^+. In the Afterthoughts to this lecture, we'll describe the 24-dimensional cases.

It is easy to evaluate the theta-functions of D_n and D_n^+. Jacobi's three *theta-constants* are defined by

$$\theta_3 = \sum q^{k^2}$$
$$\theta_4 = \sum (-1)^k q^{k^2}$$
$$\theta_2 = \sum q^{(k+\frac{1}{2})^2},$$

the summations being over all integers k, where $q = \exp(2\pi i z)$.

Now the cubic lattice I_n (or $\mathbf{Z}^n$) of all vectors $(x_1, x_2, \ldots, x_n)$ with integer x_i has theta-function θ_3^n, and θ_4^n corresponds to the same lattice with the signs of all vectors of odd norm reversed. It follows that the theta-function of D_n is $\frac{1}{2}(\theta_3^n + \theta_4^n)$.

In a similar way, we see that the theta-functions of

$$I_n + (\tfrac{1}{2}, \tfrac{1}{2}, \tfrac{1}{2}, \ldots) \text{ and } D_n + (\tfrac{1}{2}, \tfrac{1}{2}, \tfrac{1}{2}, \ldots)$$

are θ_2^n and $\frac{1}{2}\theta_2^n$, respectively. So the theta-function of D_n^+ is $\frac{1}{2}(\theta_2^n + \theta_3^n + \theta_4^n)$.

Now D_4^+ is generated by the four mutually orthogonal vectors

$$(\tfrac{1}{2}, \tfrac{1}{2}, \tfrac{1}{2}, \tfrac{1}{2}), (\tfrac{1}{2}, \tfrac{1}{2}, -\tfrac{1}{2}, -\tfrac{1}{2}), (\tfrac{1}{2}, -\tfrac{1}{2}, \tfrac{1}{2}, -\tfrac{1}{2}), (\tfrac{1}{2}, -\tfrac{1}{2}, -\tfrac{1}{2}, \tfrac{1}{2});$$

so is isomorphic to I_4, and by taking theta-functions we obtain

$$\theta_3^4 = \frac{1}{2}(\theta_2^4 + \theta_3^4 + \theta_4^4),$$

or

$$\theta_3^4 = \theta_2^4 + \theta_4^4,$$

a relation due to Jacobi.

The reader will now be able to check that the theta-functions

$$\tfrac{1}{2}(\theta_2^{16} + \theta_3^{16} + \theta_4^{16})$$

and

$$(\tfrac{1}{2}(\theta_2^8 + \theta_3^8 + \theta_4^8))^2$$

of D_{16}^+ and E_8^2 are identical, by using Jacobi's relation to eliminate θ_3 from both.

The 12-dimensional and 8-dimensional examples

Kneser [Kne] soon reduced the dimension from 16 to 12. The two lattices D_{16}^+ and $E_8 \oplus E_8$ not only have the same number of vectors of every length; they also have the same number of 2-dimensional sublattices of every possible shape. In fact, these two lattices have the same number of copies of every sublattice of dimension below 4, so that they are isospectral in a very strong sense. By taking sublattices orthogonal to copies of D_4, Kneser obtained an isospectral pair of 12-dimensional lattices, namely D_{12} and $E_8 \oplus D_4$.

Sometime later, Kitaoka[Kit] reduced the dimension even further, by producing a pair of isospectral lattices of dimension 8, each with determinant 81.

Exercise: work out the theta-functions of D_{12} and $E_8 \oplus D_4$, and use Jacobi's relation to verify their equality.

The 6-dimensional cubic and isocubic lattices

Neil Sloane is very fond of the two codes described below. The digits in these codewords are integers modulo 2 (that is, C_1 and C_2 are binary codes). C_1 and C_2 each consist of 8 codewords:

C_1	C_2
0 0 0 0 0 0	0 0 0 0 0 0
1 1 0 0 0 0	1 0 1 0 0 0
0 0 1 1 0 0	0 0 1 0 1 0
0 0 0 0 1 1	1 0 0 0 1 0
0 0 1 1 1 1	0 1 0 1 1 1
1 1 0 0 1 1	1 1 0 1 0 1
1 1 1 1 0 0	0 1 1 1 0 1
1 1 1 1 1 1	1 1 1 1 1 1

These two codes are linear: the sum of two words in either of the codes is another word in that code. Moreover, they have the same weight distribution: there is one word of weight 0, three of weight 2, three of weight 4, and one of weight 6. (The *weight* of a codeword is the number of nonzero entries in it.) However, these isospectral codes are not isomorphic because the sum of the weight-2 codewords in C_1 is 1 1 1 1 1 1, while in C_2 it is 0 0 0 0 0 0.

We can now obtain two isospectral non-isometric lattices L_1 and L_2 from these codes by defining L_i to consist of all vectors whose coordinates reduce modulo 2 to give a codeword of C_i. (What we mean by *non-isometric* lattices will be clear by the end of this section.)

There is an easy weight-preserving correspondence between C_1 and C_2—if a word of C_1 has just two coordinates of a given parity, exchange the second one with the coordinate just after it in cyclic order. One example of this is:

$$(1, 1, 0, 0, 0, 0) \rightarrow (1, 0, 1, 0, 0, 0).$$

Precisely the same rule gives a length-preserving correspondence between the vectors of L_1 and those of L_2:

$$(3, 5, 0, 2, 4, 6) \rightarrow (3, 0, 5, 2, 4, 6).$$

Vectors in which all coordinates have the same parity are left unchanged.

This correspondence shows that the two lattices L_1 and L_2 have the same number of vectors of any given length.

It is remarkable that L_1 is a rescaled version of the 6-dimensional cubic lattice! It is generated by its shortest vectors, namely

$$\begin{aligned}
\mathbf{v}_1 &= (1, 1, 0, 0, 0, 0) \\
\mathbf{v}_2 &= (1, -1, 0, 0, 0, 0) \\
\mathbf{v}_3 &= (0, 0, 1, 1, 0, 0) \\
\mathbf{v}_4 &= (0, 0, 1, -1, 0, 0) \\
\mathbf{v}_5 &= (0, 0, 0, 0, 1, 1) \\
\mathbf{v}_6 &= (0, 0, 0, 0, 1, -1)
\end{aligned}$$

and their negatives. Since $\mathbf{v}_1, \ldots, \mathbf{v}_6$ are mutually orthogonal vectors of the same length, L_1 is indeed a scaled copy of the cubic lattice. In

L_2, which we shall call the *isocubic lattice,* the shortest vectors are:

$$\mathbf{w}_1 = (1, 0, 1, 0, 0, 0)$$
$$\mathbf{w}_2 = (1, 0, -1, 0, 0, 0)$$
$$\mathbf{w}_3 = (0, 0, 1, 0, 1, 0)$$
$$\mathbf{w}_4 = (0, 0, 1, 0, -1, 0)$$
$$\mathbf{w}_5 = (1, 0, 0, 0, 1, 0)$$
$$\mathbf{w}_6 = (1, 0, 0, 0, -1, 0)$$

and their negatives.

Note that $\mathbf{w}_2 \cdot \mathbf{w}_3 = -1$, so $\mathbf{w}_2$ and $\mathbf{w}_3$ are not orthogonal, and hence L_1 is not isometric to L_2. This shows that in 6 (or more) dimensions, it is not even possible to hear whether a lattice has the same shape as a cubic lattice.

5-dimensional examples

We shall prove in the Afterthoughts to this lecture that cubicity *is* audible in 5 and fewer dimensions. However, the shape of a lattice may not be—by taking the sublattices orthogonal to the vector $(1, 1, 1, 1, 1, 1)$ in the above cubic and isocubic lattices we get two distinct 5-dimensional lattices that sound the same.

By the way, these examples are by no means isolated. It is possible to introduce freely varying parameters into their definitions and so get infinite families of isospectral lattices.

News flash! 4-dimensional examples

More recently, Schiemann [Sch1] has found some 4-dimensional examples by computer search. At the time of the original lectures, they lacked any comprehensible motivation or structure. By studying the first of them, Neil Sloane and I have found a particularly simple 4-parameter family of examples which we call the *tetralattices.* The name tetralattice is to recall the connection with the *tetracode,* the length 4 ternary code whose 9 words are:

$$(0, 0, 0, 0), \quad \pm(0, 1, 1, 1),$$

$$\pm(1, 0, 1, -1), \quad \pm(1, -1, 0, 1), \quad \pm(1, 1, -1, 0).$$

Let e_a, e_b, e_c, e_d be 4 mutually orthogonal vectors of different lengths, with

$$e_a \cdot e_a = \frac{1}{12}a, \quad e_b \cdot e_b = \frac{1}{12}b, \quad e_c \cdot e_c = \frac{1}{12}c, \quad e_d \cdot e_d = \frac{1}{12}d.$$

We use $[w, x, y, z]$ for $we_a + xe_b + ye_c + ze_d$ and define the lattice $L^+ = L^+(a, b, c, d)$ to be the lattice spanned by

$$v_1^+ = [+3, -1, -1, -1]$$
$$v_2^+ = [+1, +3, +1, -1]$$
$$v_3^+ = [+1, -1, +3, +1]$$
$$v_4^+ = [+1, +1, -1, +3].$$

Similarly, we define $L^- = L^-(a, b, c, d)$, replacing $+3$ by -3: it is spanned by

$$v_1^- = [-3, -1, -1, -1]$$
$$v_2^- = [+1, -3, +1, -1]$$
$$v_3^- = [+1, -1, -3, +1]$$
$$v_4^- = [+1, +1, -1, -3].$$

If we read any vector of either tetralattice modulo 3 we get a tetracode word. Also, any nonzero codeword of the tetracode comes from reading one of the four basis vectors or their negatives modulo 3. For instance: $v_1^+ + v_2^+ = [4, 2, 0, -2] \equiv [1, -1, 0, 1]$ (mod 3), and this tetracode word arises from reading v_3^+ (mod 3).

The kernel of this map from L^+ onto the tetracode is a sublattice M^+ of index 9. M^+ consists of vectors whose coordinates are all divisible by 3. For example,

$$v_1^+ + v_2^+ - v_3^+ = [3, 3, -3, -3] \in M^+$$

It is easy to check that M^+ is generated by the 8 vectors $[\pm 3, \pm 3, \pm 3, \pm 3]$ in which there are an even number of minus signs. In a precisely similar way, L^- contains a lattice M^- of index 9 generated by $[\pm 3, \pm 3, \pm 3, \pm 3]$ with an odd number of minus signs.

The coset representatives for M^ϵ in L^ϵ are

$$0, \pm \mathbf{v}_1^\epsilon, \pm \mathbf{v}_2^\epsilon, \pm \mathbf{v}_3^\epsilon, \pm \mathbf{v}_4^\epsilon.$$

Now M^+ is clearly isometric to M^-, since we can obtain an isometry by changing the sign of any one coordinate. Similarly, $M^+ + \mathbf{v}_i^+$ and $M^+ - \mathbf{v}_i^+$ are isometric to $M^- + \mathbf{v}_i^-$ and $M^- - \mathbf{v}_i^-$. We can combine these observations into a simple rule that gives a one-to-one length-preserving correspondence between L^+ and L^-: *change the sign of the first coordinate that is divisible by 3*. Many of Schiemann's examples are particular cases of this construction; his first pair of isospectral lattices of determinant $1729 = 7 \times 13 \times 19$ are $L^+(1, 7, 13, 19)$ and $L^-(1, 7, 13, 19)$. However, there are other 4-dimensional examples that are not in this family.

We believe that if a, b, c, d are distinct, then $L^+(a, b, c, d)$ and $L^-(a, b, c, d)$ are not isometric. This can be easily verified for the integral cases whose determinant $abcd$ is less than 10^4. For instance, we find all solutions of the equations $x^2 + 7y^2 + 13z^2 + 19w^2 = 48$ and $x^2 + 7y^2 + 13z^2 + 19w^2 = 96$ to find the possible vectors of norm 4 and 8 in $L^+(1, 7, 13, 19)$ and $L^-(1, 7, 13, 19)$. It turns out that $\pm \mathbf{v}_1^+$ and $\pm \mathbf{v}_2^+$ are the only vectors of norm 4 and 8, respectively, in $L^+(1, 7, 13, 19)$, and $\pm \mathbf{v}_1^-$ and $\pm \mathbf{v}_2^-$ are the only such vectors in $L^-(1, 7, 13, 19)$. Since $\mathbf{v}_1^+ \cdot \mathbf{v}_2^+ = -1$ but $\mathbf{v}_1^- \cdot \mathbf{v}_2^- = 2$, the two lattices cannot be isometric. See [CS2] for more details.

Late extra! Are there examples in 2 or 3 dimensions?

If we are given the lengths of all vectors in a lattice, then we can work out the lengths of its primitive vectors using the fact that the multiples

$$\pm \mathbf{v}, \pm 2\mathbf{v}, \pm 3\mathbf{v}, \pm 4\mathbf{v}, \ldots$$

of a primitive vector of norm n contribute the sum

$$2q^n + 2q^{4n} + 2q^{9n} + 2q^{16n} + \cdots$$

to the theta function. From this, one can show that if the theta function is $1 + 2\sum a_n q^n$, then the number of pairs $\pm \mathbf{v}$ of primitive vectors of norm n is

$$\sum_{d|n} \mu(d) a_{n/d^2},$$

where $\mu(d)$ is the Möbius function.

However, in the first lecture, we showed that for a 2-dimensional lattice, the three shortest primitive vectors are the values of the corresponding quadratic form at some superbase. It follows that in the 2-dimensional case, the theta function of a lattice determines its shape.

The 3-dimensional case remained open for many years. However, as this book was being written, we heard that Schiemann has shown by a large computer calculation that indeed we can hear the shape of a 3-dimensional lattice, so there is no example in this dimension either.

No, you can't hear the shape of a drum!

Although the tori obtained from distinct 2-dimensional lattices can't sound the same, there are two distinct polygonal regions that do. We shall prove this very simply, for those readers who understand the technical terms involved.

When in 1965, Mark Kac asked "can one hear the shape of a drum?", he popularized the question of whether there can exist two noncongruent isospectral domains in the plane. In the ensuing 25 years many examples of isospectral manifolds were found, whose dimensions, topology, and curvature properties gradually approached those of the plane. Recently, Gordon, Webb, and Wolpert finally got some examples in the plane. In this section, adapted from [BCDS], we give a simple pair of examples, and an easy method of proof due to Buser.

Consider the two propeller-shaped regions (a) and (b).

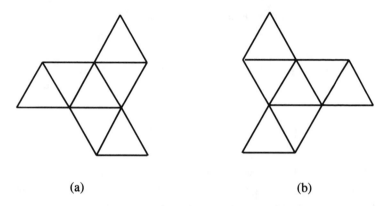

(a)   (b)

Our examples are obtained from these by replacing the equilateral triangles by acute-angled scalene triangles, all of the same shape, in such a way that any two triangles that meet along a line are mirror-images in that line, as in (c) and (d) below (drawn for the 45°–60°–75° case). In the proof that follows, we shall understand that the basic triangles are really scalene; however, we shall represent them as equilateral in order to display the symmetries of the argument, merely indicating the three lengths of edge by three different styles.

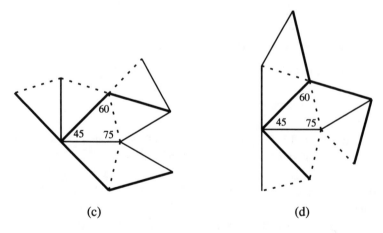

(c)   (d)

Now let λ be any real number, and $\phi = \phi(x)$ be any eigenfunction of the Laplacian with eigenvalue λ for the Dirichlet problem corresponding to the left-handed propeller shown in (e). Let a, b, c,

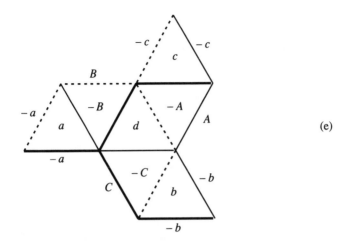

(e)

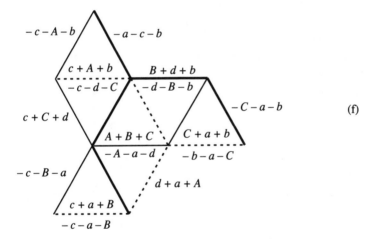

(f)

d, $-A$, $-B$, $-C$ denote the restriction of ϕ to the seven triangles of this propeller.

The Dirichlet boundary condition is that ϕ must vanish on each boundary segment. Using the reflection principle, this is equivalent to the assertion that ϕ would change sign if continued as a smooth eigenfunction across any boundary segment, as also indicated in (e).

In (f), we show how to obtain from ϕ another eigenfunction of eigenvalue λ, this time for the right-handed propeller. In the central triangle, we place the function $A(x) + B(x) + C(x)$. Now we see from (e) that the functions $A(x)$, $B(x)$, $C(x)$ continue smoothly across dashed lines into copies of the functions $-d(x)$, $-B(x)$, $-b(x)$ respectively, so that their sum continues into $-[d(x) + B(x) + b(x)]$ as shown. The reader should check in a similar way that this continues across a thick line into $B(x) + d(x) + b(x)$ (its negative), and across a thin line to $C(x) + a(x) + b(x)$, which continues across either a thick or dashed line to its own negative, as also shown in (f).

These results, and the similar ones obtained by symmetry, are enough to show that the function specified in (f) is an eigenfunction of eigenvalue λ that vanishes along each boundary section of the right-handed propeller.

We have just defined a linear map from the λ-eigenspace for the left-handed propeller to that of the right-handed one, and since we can also define an inverse map in the reverse direction, we see that these two spaces have the same dimension for all λ, proving that the two propellers are Dirichlet isospectral. They are also isospectral for the Neumann boundary condition (that the derivative normal to the boundary should vanish at the boundary), as can be seen by a similar transplantation proof in which every $-$ sign is replaced by a $+$ sign.

This type of "transplantation proof" is due to Peter Buser. Our "propellers" seem to be the simplest examples known.

What *can* we hear about a lattice?

We have said that a lattice property is *audible* if it is determined by the theta function

$$\theta_L(z) = \sum_{\mathbf{v} \in L} q^{N(\mathbf{v})}, \qquad (q = e^{2\pi i z}),$$

where $N(\mathbf{v})$ is the squared length of $\mathbf{v}$. Now geometrically, the lattice determinant d is the square of the volume of the fundamental parallelotope. It follows that the number of lattice points inside a large ball will be roughly the volume of the ball divided by $\sqrt{d}$, and so (since we can "hear" the number of $\mathbf{v}$ that have $N(\mathbf{v}) < R$ for any R):

The determinant is an audible property.

The *dual lattice* L^* is defined to be the set of all vectors in the real space spanned by L whose inner product with every vector of L is integral. There is a formula of Jacobi that relates the theta functions of L and L^*:

$$\theta_{L^*}(z) = (\det L)^{1/2} \left(\frac{i}{z}\right)^{n/2} \theta_L(-1/z),$$

and from this we deduce that

The theta function of L^* is an audible property of L.

Gauss means

There are some other very useful invariants that are audible. If f is the quadratic form corresponding to L, then we define the *Gauss mean* of f to be the mean of the numbers

$$e^{2\pi i N(\mathbf{v})}$$

over all vectors $\mathbf{v}$ of the dual lattice L^*.[1]

This is the mean of an infinite set of complex numbers, so some care is required for a formal definition in general. However, for integral lattices there is really no problem, because these numbers depend periodically on position. This happens because when L is integral, it is a sublattice of L^*, and the norms of two vectors $\mathbf{v}$ and $\mathbf{v} + \mathbf{w}$ of L^* whose difference is a vector $\mathbf{w}$ of L differ by the integer $2(\mathbf{v}, \mathbf{w}) + N(\mathbf{w})$, which entails that the terms of the Gauss mean corresponding to $\mathbf{v}$ and $\mathbf{v} + \mathbf{w}$ are the same. The Gauss mean for an integral lattice is therefore the same as the mean of

$$e^{2\pi i N(\mathbf{v})}$$

taken over a set of representative vectors $\mathbf{v}$ for the *dual quotient group* L^*/L of L, which is a finite group.

---

[1] We will modify this definition slightly in the Fourth Lecture.

The *direct sum* $f \oplus g$ of the quadratic forms f and g with matrices M and N, is the quadratic form with matrix

$$\begin{pmatrix} M & 0 \\ 0 & N \end{pmatrix}$$

It is a useful property of Gauss means that the Gauss mean of $f \oplus g$ is the product of those for f and g. This makes it easy for us to compute the Gauss mean for diagonal forms of the shape

$$\text{diag}[1, 1, \ldots, p, p, \ldots, p^2, \ldots] \qquad (p \text{ odd}).$$

(We shall omit the "diag" in such notations from now on.) By the multiplicative property of Gauss means, it will suffice to compute the Gauss mean for a 1-dimensional form $[p^k]$. Plainly the Gauss mean for $[1]$ is 1.

Now a celebrated theorem of Gauss asserts that the sum

$$\sum_{0 \leq n < p} e^{2\pi i n^2 / p}$$

has value

$$\sqrt{p} \quad \text{or} \quad i\sqrt{p},$$

according as

$$p \equiv 1 \quad \text{or} \quad p \equiv -1$$

modulo 4. Since the Gauss mean for $[p]$ is the mean of the terms in this sum, its value is correspondingly $(1 \text{ or } i)/\sqrt{p}$.

To see what happens for higher powers of p, we consider the Gauss mean for $[9]$. Since the squares of 0, 1, 2, 3, 4, 5, 6, 7, 8 are 0, 1, 4, 0, 7, 7, 0, 4, 1 modulo 9, this is the mean of

$$1, \epsilon, \epsilon^4, 1, \epsilon^7, \epsilon^7, 1, \epsilon^4, \epsilon \qquad (\epsilon = e^{2\pi i/9})$$

which is $(1 + 1 + 1)/9 = 1/3$ since

$$\epsilon + \epsilon^4 + \epsilon^7 = \epsilon(1 + \epsilon^3 + \epsilon^6) = 0.$$

In a similar way, we find that the Gauss mean for $[p^{2k}]$ is $1/p^k$, while that for $[p^{2k+1}]$ is $(1 \text{ or } i)/p^{(k+\frac{1}{2})}$, namely the mean for $[p]$ divided by p^k.

We summarize these results in the following.

Theorem. *The Gauss mean of a form f which is the direct sum of*

n copies of $[1]$, n' copies of $[p]$, n'' of $[p^2]$ ...

is

$$1^n \times \left(\frac{1}{\sqrt{p}}\right)^{n'} \times \left(\frac{1}{\sqrt{p^2}}\right)^{n''} \cdots$$

times a fourth root of unity.

Of course the displayed number is $1/\sqrt{\det(f)}$. Now let us see what happens when we divide f by a power of p. Since

$$e^{2\pi i p} = e^{2\pi i p^2} = \cdots = 1,$$

the Gauss means of

$$[1/p], \ [1/p^2], \ [1/p^3], \ldots$$

are all 1. So the Gauss mean of f/p is

$$1^{n'} \times \left(\frac{1}{\sqrt{p}}\right)^{n''} \times \left(\frac{1}{\sqrt{p^2}}\right)^{n'''} \cdots;$$

while that for f/p^2 is

$$1^{n''} \times \left(\frac{1}{\sqrt{p}}\right)^{n'''} \cdots$$

and so on, times various fourth roots of 1. Since all these numbers are audible, we conclude

Theorem. *If p is odd, then for forms of the shape $[1, 1, \ldots, p, p, \ldots, p^2, \ldots, p^k]$ the numbers $n, n', n'', \ldots$ of terms of each value are all audible.*

We shall return to this subject for more general forms after the Fourth Lecture. The numbers $n, n', n'' \ldots$ are the exponents in the p-adic symbol defined there, and from the fourth roots of unity we shall also derive the signs that occur in that symbol. The p-adic symbol of that Lecture will prove to be an audible property for odd p, but not for $p = 2$.

Kneser's Gluing Method:
Unimodular Lattices

Introduction

The main topic of this book is classification of quadratic forms—this Lecture has been a digression whose relevance will only become apparent later. The First Lecture classified 2-dimensional forms, the Third will classify definite 3-dimensional forms, and the Fourth will classify indefinite forms in all dimensions greater than 2.

There is no hope of classifying positive definite quadratic forms in high dimensions. However, Kneser obtains many integral lattices of small determinant by "gluing" root lattices to each other (or themselves). Milnor's toroidal "drums" used the 16-dimensional even unimodular lattices E_8^2 and D_{16}^+. A more spectacular application was Niemeier's enumeration of all the 24-dimensional even unimodular lattices.

An example of glue

We shall describe this *gluing method* by means of a simple example. Let R, S be two orthogonal 1-dimensional lattices in the plane generated by vectors $\mathbf{r}$ and $\mathbf{s}$, respectively, for which $N(\mathbf{r}) = \mathbf{r} \cdot \mathbf{r} = 2$, $N(\mathbf{s}) = \mathbf{s} \cdot \mathbf{s} = 2$, and $\mathbf{r} \cdot \mathbf{s} = 0$.

We ask for a 2-dimensional integral lattice L that properly contains $R \oplus S$. Any vector $\mathbf{y}$ of L must have the form $\mathbf{y} = \lambda\mathbf{r} + \mu\mathbf{s}$, where $\mathbf{y} \cdot \mathbf{r} = \lambda\mathbf{r} \cdot \mathbf{r} = 2\lambda$ and $\mathbf{y} \cdot \mathbf{s} = \mu\mathbf{s} \cdot \mathbf{s} = 2\mu$ must be integers, so

λ must be half of an integer, and similarly for μ. Another restriction is that $\mathbf{y} \cdot \mathbf{y} = 2\lambda^2 + 2\mu^2$ be an integer. There's really only one nontrivial possibility modulo $R \oplus S$, namely $\mathbf{y} = \frac{1}{2}\mathbf{r} + \frac{1}{2}\mathbf{s}$. We say that the lattice L generated by this $\mathbf{y}$ together with R and S is obtained by *gluing* R and S by the glue vector $\mathbf{y}$. In the figure, circles indicate points of $R \oplus S$, and crosses the remaining points of L.

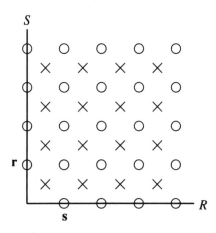

The same method can be used to glue any number of integral lattices, say $L_1, L_2, \ldots, L_k$ together to form an integral lattice L containing $L_1 \oplus L_2 \ldots \oplus L_k$. Then L is generated by $L_1, L_2, \ldots, L_k$ along with various glue vectors, which can be written $\mathbf{y}_1 + \mathbf{y}_2 \ldots + \mathbf{y}_k$ where $\mathbf{y}_i$ is in the real space spanned by L_i and has integral inner product with every vector of L_i. Also, the value of $\mathbf{y}_i$ is only important modulo L_i. The possible $\mathbf{y}_i$, which correspond to the elements of the dual quotient group L_i^*/L_i, will be called the *glue vectors* for L_i.

Root lattices

There is a class of lattices called *root lattices,* which it's particularly nice to glue together. We briefly describe them and their glue vectors.

The lattice A_n consists of all vectors $(x_0, x_1, \ldots, x_n)$ whose coordinates are $n + 1$ integers with zero sum. For instance A_2 is the 2-

dimensional lattice spanned by $(1, -1, 0)$ and $(0, 1, -1)$. Its glue vectors will therefore have the form (x, y, z) where $x + y + z = 0$, but x, y, z need not be integers. One such possibility is $(2/3, -1/3, -1/3)$. For general n, the non-zero glue vectors of A_n are

$$[i] = \left(\left(\frac{j}{n+1} \right)^i, \left(\frac{-i}{n+1} \right)^j \right) \qquad 1 \le i \le n, \quad i + j = n + 1.$$

The dual quotient group A_n^*/A_n is cyclic of order $n + 1$.

We have already met the root lattice D_n, consisting of all vectors $(x_1, \ldots, x_n)$ of n integers with even sum. (This may be called the "checkerboard" lattice, since it has one vector for each black "cell" of the n-dimensional analogue of a checkerboard.) Its nonzero glue vectors are

$$[1] = \left(\tfrac{1}{2}, \tfrac{1}{2}, \ldots, \tfrac{1}{2} \right),$$
$$[2] = (0, \ldots, 0, 1),$$
$$[3] = \left(\tfrac{1}{2}, \tfrac{1}{2}, \ldots, \tfrac{1}{2}, -\tfrac{1}{2} \right).$$

The dual quotient group D_n^*/D_n has order 4, being cyclic just if n is odd.

The lattice E_8 consists of all vectors $(x_1, \ldots, x_8)$ whose eight coordinates are either all integers or all half-integers and have even sum. There are no nonzero glue vectors for E_8, so the dual quotient E_8^*/E_8 is trivial.

The lattice E_7 consists of all vectors in E_8 that have $x_1 + \cdots + x_8 = 0$. The nonzero glue vector is

$$[1] = \left(\tfrac{1}{4}, \ldots, \tfrac{1}{4}, -\tfrac{3}{4}, -\tfrac{3}{4} \right),$$

and E_7^*/E_7 has order 2.

The lattice E_6 consists of all vectors in E_8 that have $x_1 + x_8 = x_2 + \cdots + x_7 = 0$. The nonzero glue vectors are

$$[1] = \left(0, -\tfrac{2}{3}, -\tfrac{2}{3}, \tfrac{1}{3}, \tfrac{1}{3}, \tfrac{1}{3}, \tfrac{1}{3}, 0 \right) \quad \text{and} \quad [2] = -[1],$$

and E_6^*/E_6 has order 3.

Of course, the zero vector $[0]$ is a glue vector in every case.

Gluing lattices together

We shall introduce a convenient notation for the lattices obtained by gluing root lattices together. Suppose that $X_m, Y_n, \ldots$ are root lattices of dimensions $m, n, \ldots$. Then we'll let

$$X_m Y_n \ldots [ab\ldots, \quad a'b'\ldots, \quad \ldots]$$

denote the lattice obtained by adjoining the glue vectors

$$[a] + [b] + \cdots,$$
$$[a'] + [b'] + \cdots,$$

$$\cdots$$

to $X_m \oplus Y_n \oplus \cdots$.

For example, one of Niemeier's 24-dimensional lattices is denoted $A_{11} D_7 E_6 [1,1,1]$. Now

$$[1] \text{ for } A_{11} \text{ is } (\tfrac{11}{12}, \tfrac{-1}{12}, \ldots, \tfrac{-1}{12}),$$
$$[1] \text{ for } D_7 \text{ is } (\tfrac{1}{2}, \ldots, \tfrac{1}{2}),$$
$$[1] \text{ for } E_6 \text{ is } (0, -\tfrac{2}{3}, -\tfrac{2}{3}, \tfrac{1}{3}, \tfrac{1}{3}, \tfrac{1}{3}, \tfrac{1}{3}, 0).$$

So $A_{11} D_7 E_6 [1,1,1]$ is the lattice spanned by $A_{11} \oplus D_7 \oplus E_6$ together with the vector

$$\left(\tfrac{11}{12}, (\tfrac{-1}{12})^{11}, \ (\tfrac{1}{2})^7, \ 0, -\tfrac{2}{3}, -\tfrac{2}{3}, (\tfrac{1}{3})^4, 0\right).$$

We can simplify this notation to $(A_{11} D_7 E_6)^+$ on occasions when it is not necessary to specify the nontrivial glue vectors more precisely. The final $+$ just means "together with some glue vectors".

Niemeier lattices

Many interesting lattices can be described in this manner. In 8 dimensions, there is only one even unimodular lattice, namely E_8. As we saw in the second lecture, there are two such lattices D_{16}^+ and E_8^2, in 16 dimensions. In this notation they are $D_{16}[1]$ and $E_8 E_8 [0,0]$ respectively. In general, we shall use the notation $(X_m Y_n \cdots)^+$ to denote the result of adjoining some unspecified glue vectors to $X_m \oplus Y_n \oplus \cdots$

Niemeier solved an old problem of Witt by enumerating the possible even unimodular 24-dimensional lattices. They are described in the table below. Parentheses in a glue vector there mean that all vectors resulting from cyclic permutations of that part of the vector are also to be used as glue vectors. The last lattice, Λ_{24}, is the remarkable lattice discovered by John Leech, which we unfortunately cannot discuss here.

Lattice	Glue (omitting commas)
$(D_{24})^+$	[1]
$(D_{16}E_8)^+$	[10]
E_8^3	[000]
$(A_{24})^+$	[5]
$(D_1 2^2)^+$	[12, 21]
$(A_{17}E_7)^+$	[31]
$(D_{10}E_7^2)^+$	[110, 301]
$(A_{15}D_9)^+$	[21]
$(D_8^3)^+$	[(122)]
$(A_{12}^2)^+$	[15]
$(A_{11}D_7E_6)^+$	[11]
$(E_6^4)^+$	[1(012)]
$(A_9^2D_6)^+$	[240, 501, 053]
$(D_6^4)^+$	[even perms of (0123)]
$(A_8^3)^+$	[(114)]
$(A_7^2D_5^2)^+$	[1112, 1721]
$(A_6^4)^+$	[1(216)]
$(A_5^4D_4)^+$	[(2(024)0, 33001, 30302, 30033]
$(D_4^6)^+$	[111111, 0(02332)]
$(A_4^6)^+$	[1(01441)]
$(A_3^8)^+$	[3(2001011)]
$(A_2^{12})^+$	[2(11211122212)]
$(A_1^{24})^+$	[1(00000101001100110101111)]
Λ_{24}	(not obtainable by this method)

The same method can be used to enumerate other lattices. For instance, all the unimodular lattices of dimension at most 16 are direct sums of copies of

$$I_1, \ E_8, \ D_{12}[1], \ E_7^2[11], \ A_{15}[4], \ D_8^2[(12)], \ D_{16}[1],$$

where I_1 is the 1-dimensional lattice generated by a vector of norm 1.

Witt's lemma on root lattices

To further illuminate the utility of glue, we prove a pleasant lemma of Witt which asserts that the integral lattices generated by norm-2 vectors are direct sums of root lattices A_n, D_n, E_n.

It will suffice to prove that an indecomposable lattice generated by norm-2 vectors is of the form A_n, D_n, or E_n. Suppose that L is an indecomposable counterexample, and consider the maximal root lattice X_n contained in L. Then there is a norm-2 vector of L not in X_n, and indeed there is such a vector not orthogonal to X_n (or else L would split). This vector $\mathbf{v}$ equals $\mathbf{v}_1 + \mathbf{v}_2$ where $\mathbf{v}_1 \neq \mathbf{0}$ is in the space of X_n ("is horizontal"), $\mathbf{v}_2$ is orthogonal to that space ("is vertical"), and $N(\mathbf{v}_1) + N(\mathbf{v}_2) = 2$. Also $\mathbf{v}_1$ must have integral inner product with every vector in X_n, since $\mathbf{v}_2$ contributes nothing to such inner products.

So we are in a position to use gluing theory. Here $\mathbf{v}_1$ has to be a glue vector of X_n and we are gluing X_n to the lattice generated by $\mathbf{v}_2$. In each case we will find that the resulting lattice is also a root lattice, contradicting the maximality.

As an example, we consider a lattice generated by A_9 together with some $\mathbf{v} = \mathbf{v}_1 + \mathbf{v}_2$ of norm 2. The vector $\mathbf{v}_1$ must be a glue vector for A_9 of norm at most 2. The only possibilities are $\mathbf{v}_1 = [1]$ or $[9]$ of norm $9/10$, glued to $\mathbf{v}_2$ of norm $11/10$ and $\mathbf{v}_1 = [2]$ or $[8]$ of norm $16/10$, glued to $\mathbf{v}_2$ of norm $4/10$. So there are just two such lattices, which must therefore be A_{10} and D_{10}, since these do both contain A_9.

If, more generally, X_n extends like this to an $n + 1$-dimensional lattice Y_{n+1}, the ratio of the determinants of these lattices will be $N(\mathbf{v}_2) = 2 - N(\mathbf{v}_1)$. Some cases that we know must arise are:

	$N(\mathbf{v}_2)$	$N(\mathbf{v}_1)$	Glue
A_{n+1} from A_n	$(n+2)/(n+1)$	$n/(n+1)$	[1] or [n]
D_{n+1} from D_n	1	1	[2]
E_6 from A_5	$3/6 = 1/2$	$3/2$	[3]
E_6 from D_5	$3/4$	$5/4$	[1] or [3]
E_7 from A_6	$2/7$	$12/7$	[3] or [4]
E_7 from D_6	$2/4 = 1/2$	$3/2$	[1] or [3]
E_7 from E_6	$2/3$	$4/3$	[1] or [2]
E_8 from A_7	$1/8$	$15/8$	[3] or [5]
E_8 from D_7	$1/4$	$7/4$	[1] or [3]
E_8 from E_7	$1/2$	$3/2$	[1]

The glue vector $\mathbf{v}_1$ that we must have used has known norm h, and we have indicated the possibilities for the glue vector $\mathbf{v}_1$ (whose norm is now known) at the end of the lines. The list is complete because one can check that it includes *all* glue vectors of norm less than 2.

If instead we extend X_n to another n-dimensional lattice Y_n, then $N(\mathbf{v}_2)$ will be 0 and $N(\mathbf{v}_1)$ will be 2. Some cases that we know must happen are:

	$N(\mathbf{v}_2)$	$N(\mathbf{v}_1)$	Glue
E_7 from A_7	0	2	[4]
E_8 from A_8	0	2	[3] or [6]
E_8 from D_8	0	2	[1] or [3].

Again the glue vector $\mathbf{v}_1$ is determined well enough by its norm of 2, and all cases of norm-2 glue vectors have arisen.

Since every integral lattice generated by norm-2 vectors must arise by successive extensions of this type using glue vectors of norm at most 2, and since each such glue vector arises in one of the extensions that gives a root lattice, *every* such extension must give a root lattice.

Of course it is Witt's lemma that makes the root lattices so useful in constructing other integral lattices of small determinant. The point is that such a lattice is likely to have many vectors of norm 1 or 2, which will generate the direct sum of a cubic lattice I_n and some of these root lattices.

Cubicity is audible in low dimensions

We can use Witt's lemma to prove the assertion of the text that cubicity is an audible property in dimensions up to 5.

Rescale the 5-dimensional cubic lattice to the lattice

$$L = A_1 \oplus A_1 \oplus A_1 \oplus A_1 \oplus A_1,$$

whose vectors all have even integral norm. A lattice L' that sounds the same as L will have the same property, and so be an even integral lattice containing 10 vectors $\pm \mathbf{v}_1, \pm \mathbf{v}_2, \pm \mathbf{v}_3, \pm \mathbf{v}_4, \pm \mathbf{v}_5$ of norm 2. Witt's lemma tells us that these generate a sublattice of L' that is a direct sum of root lattices, which can only be

$$A_1 \oplus A_1 \oplus A_1 \oplus A_1 \oplus A_1$$

again (when $L' = L$), or

$$A_2 \oplus A_1 \oplus A_1.$$

So if $L' \neq L$, it can be obtained by gluing some lattice of the form

$$L'' = A_2 \oplus A_1 \oplus A_1 \oplus L_1$$

where L_1 is the 1-dimensional lattice generated by the shortest vector of L' that is orthogonal to $A_2 \oplus A_1 \oplus A_1$ (we suppose that this vector has norm n).

Now the determinant of L'' is $3 \times 2 \times 2 \times n$, which must be $32d^2$, where d is the index of L'' in L'. We see that d is divisible by 3, so that L' must contain a glue vector of order 3 modulo L''. But up to sign, the only glue vector of order 3 for $A_2 \oplus A_1 \oplus A_1$ is $[1] \oplus [0] \oplus [0]$, of norm $2/3$, and the only one for L_1 is $v/3$, of norm $n/9$. So $2/3 + n/9$ must be an integer, which contradicts the above assertion that $12n = 32d^2$.

... and Can You Feel Its Form?

Geometry or Arithmetic?

When we discussed binary quadratic forms in the First Lecture, there was a marked difference between the definite and indefinite cases. This persists in higher dimensions. The values of a positive definite form are best regarded as squared lengths of vectors in a lattice, and we classify such forms by discussing the shape of this lattice geometrically.

In the indefinite case, when the dimension is at least 3, there is a complete classification, due to Eichler, in terms of an arithmetical invariant called the spinor genus, which is defined in terms of a simpler and more important invariant, the genus.

In my Hedrick Lectures, I compressed these two very different topics into one session. In print it has seemed better to separate them. This Third Lecture mainly concerns the geometrical classification of 3-dimensional lattices in terms of the shape of their Voronoi cells. The arithmetical discussion is postponed until the Fourth Lecture, after which we shall describe the Eichler theorem.

The Voronoi cell

We recall from the first lecture that positive definite binary forms can be specified either by three numbers α, β, γ, called the conorms, or three other numbers a, b, c, called the vonorms. We shall now

61

interpret all these numbers geometrically and generalize them to higher dimensions.

There are two facts to keep in mind for this discussion. First, we have rules for converting between Greek and Roman, namely $a = \beta + \gamma$, $b = \alpha + \gamma$, and $c = \alpha + \beta$, and so $\alpha = \frac{1}{2}(b + c - a)$, etc. Also, the superbase e_1, e_2, e_3 of the well has the property that $e_i \cdot e_j \leq 0$; we call this an *obtuse superbase*. We'll obtain the corresponding results for 3-dimensional lattices.

The geometrical interpretation is that the vonorms are in general the norms of certain vectors called the Voronoi vectors. In 2 or 3 dimensions, the conorms are essentially the same as certain parameters introduced by Selling in 1874, namely the negatives of the inner products of pairs of vectors from an obtuse superbase. The Voronoi vectors are associated with an important region called the Voronoi cell (or Dirichlet cell, or Brillouin zone).

To each point of the lattice, attach the set of all points in the space that are at least as near to that point as to any other lattice point. In this way, we obtain a tiling of the space into regions called *Voronoi cells*. Here's a picture for the Voronoi cell for the hexagonal lattice.

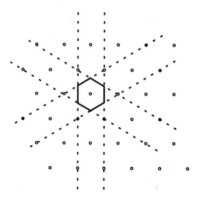

The Voronoi cell of the origin is found by a series of contests between the origin and every other lattice point v; each contest is settled by the hyperplane that perpendicularly bisects the line segment between 0 and v; the origin wins in one of the resulting half-spaces.

Then the Voronoi cell is the intersection of all these half-spaces

$$H(\mathbf{v}) = \{\mathbf{x} | \mathbf{x} \cdot \mathbf{v} \le \tfrac{1}{2}\mathbf{v} \cdot \mathbf{v}\}.$$

Now most of these hyperplanes are irrelevant in that they aren't needed to define the Voronoi cell because they're too far away from the origin. The relevant ones come from certain vectors, which we call *strict Voronoi vectors*. There is a lovely theorem of Voronoi that says: $\mathbf{v}$ is a Voronoi vector for L if and only if $\pm\mathbf{v}$ are exactly all the shortest vectors in their coset of $2L$. For if $\mathbf{w}$ is another vector in the coset of $\mathbf{v}$ and no longer than $\mathbf{v}$, then $H(\mathbf{v})$ is irrelevant, since it contains the intersection of $H(\tfrac{1}{2}(\mathbf{v} + \mathbf{w}))$ and $H(\tfrac{1}{2}(\mathbf{v} - \mathbf{w}))$. Also, if $H(\mathbf{v})$ is irrelevant, it must be because $\tfrac{1}{2}\mathbf{v}$ is outside or on some $H(\mathbf{x})$, that is, $\tfrac{1}{2}\mathbf{v} \cdot \mathbf{x} \ge \tfrac{1}{2}\mathbf{x} \cdot \mathbf{x}$, so $N(\mathbf{v} - 2\mathbf{x}) \le N(\mathbf{v})$.

The lattice $2L$ consists of the doubles of vectors of L. L has, in the 2-dimensional case, four cosets in $2L$, which we represent by four colors $\bigcirc$, $\triangle$, $\otimes$, $*$, the rule being that two vectors are given the same color exactly when their difference is twice a vector of L. Then the quotient group $L/2L$ is exactly the set of colors $\{\bigcirc, \triangle, \otimes, *\}$, made into a group.

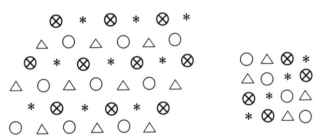

L/2L and its group structure

2-dimensional Voronoi cells

Most 2-dimensional lattices have hexagonal Voronoi cells with six Voronoi vectors $\pm\mathbf{u}$, $\pm\mathbf{v}$, $\pm\mathbf{w}$ that satisfy $\mathbf{u} + \mathbf{v} + \mathbf{w} = 0$ and look like a slight distortion of this:

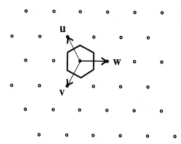

In such a case, the vonorms a, b, c are just the norms of **u**, **v**, and **w**:

$$\mathbf{u} \cdot \mathbf{u} = a, \quad \mathbf{v} \cdot \mathbf{v} = b, \quad \mathbf{w} \cdot \mathbf{w} = c,$$

while the conorms are the negatives of their inner products:

$$\mathbf{v} \cdot \mathbf{w} = -\alpha, \quad \mathbf{w} \cdot \mathbf{u} = -\beta, \quad \mathbf{u} \cdot \mathbf{v} = -\gamma.$$

Now suppose we have a rectangular lattice, with sides parallel to vectors **u** and **v**. Then there are just four Voronoi vectors, $\pm\mathbf{u}$ and $\pm\mathbf{v}$, since these are the shortest vectors in their cosets, but the remaining coset has four vectors all of the same length, ($\mathbf{u} + \mathbf{v}$, $\mathbf{u} - \mathbf{v}$ and their negatives). As we can see in the diagram, the hyperplanes for these vectors are not needed to define the Voronoi cell because they pass obliquely through the corners. So $\mathbf{u} + \mathbf{v}$ and $\mathbf{u} - \mathbf{v}$ aren't strictly Voronoi vectors; but since they were almost needed, we'll call them (and their negatives *lax Voronoi vectors.*

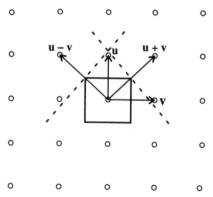

Vonorms

Let L be any lattice. Then the *vonorm* of a coset of L in $2L$ is the norm of the shortest vector in that coset. So the vonorms (Voronoi norms) are the norms of the Voronoi vectors, strict and lax. There is just one vonorm per color. We consider the 0 vonorm (from the trivial coset) to be an "improper" vonorm and so we don't usually count it. For an n-dimensional lattice, there are $2^n - 1$ proper vonorms.

Now we review what happened in the first lecture: A positive definite binary quadratic form, and hence a 2-dimensional latttice, has either a simple well or a double well. A simple well corresponds to a hexagonal Voronoi cell with a "strictly obtuse" superbase $\pm \mathbf{e}_1$, $\pm \mathbf{e}_2$, $\pm \mathbf{e}_3$, that is, $\mathbf{e}_i \cdot \mathbf{e}_j < 0$ for $i \neq j$. Here the $2^2 - 1 = 3$ vonorms are $N(\mathbf{e}_1)$, $N(\mathbf{e}_2)$, and $N(\mathbf{e}_3)$, while the conorms are $-\mathbf{e}_2 \cdot \mathbf{e}_3$, $-\mathbf{e}_3 \cdot \mathbf{e}_1$, and $-\mathbf{e}_1 \cdot \mathbf{e}_2$.

A double well corresponds to a rectangular Voronoi cell, with a base $\pm \mathbf{e}_1$, $\pm \mathbf{e}_2$ of vectors at right angles. So we really have two "laxly obtuse" superbases,

$$\{\pm \mathbf{e}_1, \pm \mathbf{e}_2, \pm(\mathbf{e}_1 + \mathbf{e}_2)\} \quad \text{and} \quad \{\pm \mathbf{e}_1, \pm \mathbf{e}_2, \pm(\mathbf{e}_1 - \mathbf{e}_2)\}.$$

In this case, the 3 vonorms are

$$a = N(\mathbf{e}_1), \quad b = N(\mathbf{e}_2) \quad \text{and} \quad a + b = N(\mathbf{e}_1 \pm \mathbf{e}_2),$$

while the conorms are b, a, and 0.

Characters and conorms

What about the mysterious relations

$$\alpha = \frac{b + c - a}{2}?$$

The conorms in n dimensions are certain numbers computed from the vonorms by a generalization of this relation involving the *real characters* of the lattice. Such a character is a map $\chi : L \to \{\pm 1\}$ with the property that $\chi(\mathbf{v} + \mathbf{w}) = \chi(\mathbf{v}) \cdot \chi(\mathbf{w})$. Usually we abbreviate the values to $+$ and $-$. Also we will neglect the "improper" character whose values are all $+$.

In the case of a 2-dimensional lattice there are three proper characters as in the table below. Vonorm space consists of the "colors" $\{\bigcirc, \triangle, \bigotimes, *\}$. Thus the vonorm function vo (color) is really defined on vonorm space and takes the values 0, a, b, c. Usually we neglect the "improper color" $\bigcirc$.

The table below illustrates the computation of conorms ("conjugate norms"), one per proper character. The *conorm* for a given character χ is defined to be

$$-\frac{1}{2^{n-1}} \sum \chi(\text{color}) vo(\text{color}).^2$$

For a 2-dimensional lattice, we have 4 colors, so 4 vonorms (3 of them proper), and 3 proper characters and conorms, as in the table.

colors: $\bigcirc \;\; \triangle \;\; \bigotimes \;\; *$

vonorms: $0 \quad a \quad b \quad c$

$$\left.\begin{array}{l}\text{characters} \\ \text{and} \\ \text{conorms}\end{array}\right\} : \left\{\begin{array}{l} + \; + \; - \; - \;\; \to \;\; \alpha = -\frac{1}{2}(a - b - c) \\ + \; - \; + \; - \;\; \to \;\; \beta = -\frac{1}{2}(b - c - a) \\ + \; - \; - \; + \;\; \to \;\; \gamma = -\frac{1}{2}(c - a - b) \end{array}\right.$$

Vonorm space and conorm space

In the 2-dimensional case, vonorms are defined on the "vonorm line" consisting of three points:

$$a\text{———}b\text{———}c$$

The points of this line correspond to the three nontrivial elements of $L/2L$.

Conorms are defined on a "dual" line called the *conorm line*:

$$\alpha\text{———}\beta\text{———}\gamma$$

The points on this line correspond to the three nontrivial characters of $L/2L$.

---

[2] In other words, the conorms—up to a constant factor—are the "finite Fourier transforms" of the vonorms.

The Fano plane: preliminaries for the 3-dimensional case

In the 3-dimensional case, $L/2L$ is a group (isomorphic to $\mathbf{Z}/2 \times \mathbf{Z}/2 \times \mathbf{Z}/2$) of order 8. So we have eight colors, seven of which are proper. The seven colors of vonorm space form a projective plane. This is a copy of the so-called Fano plane, which has seven "points" and seven "lines" of three points each, with the property that any two points lie on just one line. We may draw it thus:

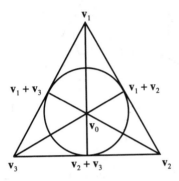

There is one vonorm for each **point** of this plane. There is also one proper character for each **line** of this plane, which takes the value $+$ for the three points on the line (and 0), and $-$ on the four points off the line. So the space on which the conorms are defined is actually the dual projective plane.

We briefly discuss the duality relation for the Fano plane. Label the points A, B, C, D, E, F, G.

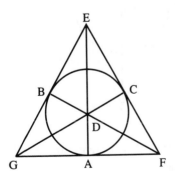

Then mark the lines a, b, c, d, e, f, g. For instance, f is the line through G, B, and E.

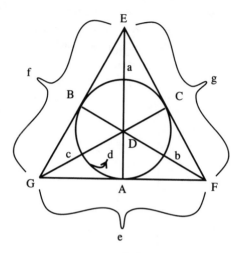

Now make a new Fano plane with points marked a, b, c, d, e, f, g, where three points are collinear just if the three corresponding lines of the original plane intersect.

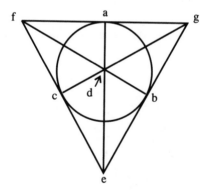

Three concurrent lines in this diagram correspond to three collinear points in the original diagram.

Vonorms and conorms for 3-dimensional lattices

A 3-dimensional lattice has eight colors O, A, B, C, D, E, F, G. So the proper vonorms $vo(A)$, $vo(B)$ ..., $vo(G)$, are the norms of the shortest vectors of the respective colors. There are eight characters χ_0, χ_a, χ_b ..., χ_g, and the seven proper ones correspond to the lines of the vonorm plane (and thus to the seven points of the conorm plane). So for example, χ_a is the character which is $+$ on O and the points A, D, and E of the line a, and $-$ on the other four points.

In 3 dimensions, the conorm of a character of χ is given by the formula

$$co(\chi) = -\frac{1}{4} \sum_{\text{colors}} \chi(\text{color}) vo(\text{color}).$$

In other words, the conorm corresponding to a given line is

$$\frac{1}{4} \left(\begin{array}{l} \text{sum of the vonorms of points off the line} \\ \text{minus sum of those of points on the line} \end{array} \right)$$

There is also a formula for recovering the vonorms from the conorms:

$$vo(\text{color}) = \sum_{\chi(\text{color})=-1} co(\chi),$$

the sum of the conorms of all characters that take the given color to -1 (or geometrically, of all lines not passing through the corresponding point of the vonorm plane).

Obtuse superbases

It turns out that, just as in the 2-dimensional case of the first lecture, every 3-dimensional lattice has at least one obtuse superbase, and in general only one. In n dimensions, an obtuse superbase (if it exists) consists of $n + 1$ vectors $\mathbf{v}_0, \mathbf{v}_1, \ldots, \mathbf{v}_n$, which have sum equal to $\mathbf{0}$ and $\mathbf{v}_i \cdot \mathbf{v}_j \leq 0$ for $i \neq j$. Now set $p_{ij} = -\mathbf{v}_i \cdot \mathbf{v}_j$. Then Selling's formula again holds: the norm of a given vector

$$\mathbf{v} = m_0\mathbf{v}_0 + m_1\mathbf{v}_1 + \cdots + m_n\mathbf{v}_n$$

is

$$N(\mathbf{v}) = \sum_{0 \le i < j \le n} p_{ij}(m_i - m_j)^2.$$

If we change the m_i's by even integers, we only alter $\mathbf{v}$ by a member of $2L$. So, when we search for the shortest vectors of each coset, we can replace all even m_i by 0 and all odd m_j by 1. So the Voronoi vectors are precisely those with $m_i = 0$ or 1, namely (in the 3-dimensional case):

$$\pm\mathbf{v}_1, \pm\mathbf{v}_2, \pm\mathbf{v}_3, \pm\mathbf{v}_0, \pm(\mathbf{v}_1 + \mathbf{v}_2), \pm(\mathbf{v}_2 + \mathbf{v}_3), \pm(\mathbf{v}_1 + \mathbf{v}_3).$$

Note that, for example, $\mathbf{v}_0 + \mathbf{v}_3 = -(\mathbf{v}_1 + \mathbf{v}_2)$. Here we have a pair of representative vectors for each of the seven nontrivial colors in vonorm space.

Now the norm of $\mathbf{v}_i$ is $p_{ij} + p_{ik} + p_{il}$, which we abbreviate to $p_{i|jkl}$, where j, k, and l are the other three indices. Also, the norm of $\mathbf{v}_i + \mathbf{v}_j$ is $p_{ik} + p_{il} + p_{jk} + p_{jl}$, abbreviated to $p_{ij|kl}$. So they appear in vonorm space as in this figure:

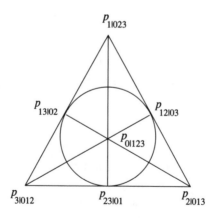

We shall show that when there is an obtuse superbase with inner products $-p_{ij}$, then the conorms are just the p_{ij}, together with 0. In the 3-dimensional case, we have labeled them in the dual Fano plane below.

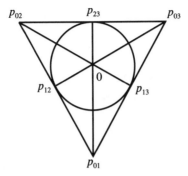

For example, the conorm corresponding to the line f that passes through $\mathbf{v}_1$ and $\mathbf{v}_3$ is

$$co(\chi_f) = -\frac{1}{4} \sum_{\text{colors}} \chi_f(\text{color})vo(\text{color})$$

$$= -\frac{1}{4}\Big(N(\mathbf{v}_1 + \mathbf{v}_3) + N(\mathbf{v}_3) + N(\mathbf{v}_1)$$

$$- N(\mathbf{v}_1 + \mathbf{v}_2) - N(\mathbf{v}_0) - N(\mathbf{v}_2 + \mathbf{v}_3) - N(\mathbf{v}_2)\Big)$$

$$= -\frac{1}{4}\big(p_{13|02} + p_{3|012} + p_{1|023}$$

$$- p_{12|03} - p_{0|123} - p_{01|23} - p_{2|013}\big)$$

$$= -\frac{1}{4}(-4p_{02}) = p_{02}.$$

A similar calculation shows the middle conorm to be 0.

Obtuse superbases in 3 dimensions

We shall prove that every 3-dimensional lattice has an obtuse superbase. For the above superbase to be obtuse, all of the p_{ij} must be positive. Now let's take L and deform it continuously until exactly one of the p_{ij} just becomes negative, say $p_{13} = -\epsilon$.

Now $p_{13} < 0$ tells us that $\mathbf{v}_1 - \mathbf{v}_3$ is now shorter than $\mathbf{v}_1 + \mathbf{v}_3$; since the other p_{ij} are positive, we can check using Selling's formula that the other six vectors are still the shortest vectors of their colors.

So we define a new, "adjacent", superbase

$$\mathbf{v}_0' = \mathbf{v}_0 + \mathbf{v}_1,$$
$$\mathbf{v}_1' = -\mathbf{v}_1,$$
$$\mathbf{v}_2' = \mathbf{v}_1 + \mathbf{v}_2,$$
$$\mathbf{v}_3' = \mathbf{v}_3.$$

It will turn out that this is an obtuse superbase for ϵ so small that $p_{ij} - \epsilon$ will be nonnegative. The resulting Voronoi vectors are

$$\mathbf{v}_0' = \mathbf{v}_0 + \mathbf{v}_1 = -(\mathbf{v}_2 + \mathbf{v}_3),$$
$$\mathbf{v}_1' = -\mathbf{v}_1,$$
$$\mathbf{v}_2' = \mathbf{v}_1 + \mathbf{v}_2,$$
$$\mathbf{v}_3' = \mathbf{v}_3,$$
$$\mathbf{v}_1' + \mathbf{v}_2' = \mathbf{v}_2,$$
$$\mathbf{v}_2' + \mathbf{v}_3' = (\mathbf{v}_1 + \mathbf{v}_2 + \mathbf{v}_3) = -\mathbf{v}_0,$$
$$\mathbf{v}_1' + \mathbf{v}_3' = -(\mathbf{v}_1 - \mathbf{v}_3).$$

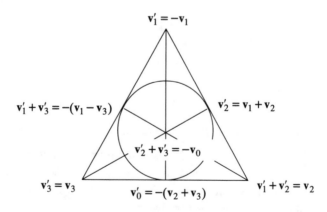

So the term "adjacent" fits, because six of these Voronoi vectors are the same as the old ones, up to sign. It is interesting to think of the obtuse superbase as like a tightened rubber band; as we deform the lattice past a critical configuration, the rubber band "snaps back" into an adjacent superbase.

Since only one vonorm $N(\mathbf{v}_1 - \mathbf{v}_3) = N(\mathbf{v}_1 + \mathbf{v}_3) - 4\epsilon$ has changed (by 4ϵ), the new conorms are found in the following way: we add ϵ to the numbers on the line going through p_{13} and 0, and we subtract ϵ from the numbers not on this line. For instance, the "upper left" conorm below is

$$-\frac{1}{4}\big(N(\mathbf{v}_1 - \mathbf{v}_3) + N(\mathbf{v}_3) + N(-\mathbf{v}_1)$$
$$- N(\mathbf{v}_1 + \mathbf{v}_2) - N(\mathbf{v}_0) - N(-(\mathbf{v}_2 + \mathbf{v}_3)) - N(\mathbf{v}_2)\big)$$

and since $N(\mathbf{v}_1 - \mathbf{v}_3) = N(\mathbf{v}_1 + \mathbf{v}_3) - 4\epsilon$, this is $p_{02} + \epsilon$.

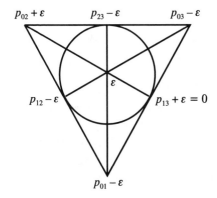

This proves that every 3-dimensional lattice L has an obtuse superbase. For let L_0 be a lattice with an obtuse superbase. Now deform it towards L until just one of the p_{ij} becomes negative, and snap back to the adjacent obtuse superbase. We continue this until we have deformed L_0 into L. Since at every stage we have an obtuse superbase, this process yields an obtuse superbase for L.

An example

In practice, we work backwards from a non-obtuse superbase to an obtuse one, tracking the inner products of the superbase vectors as it changes. Actually we work with the negatives of the inner products, since when we get to an obtuse superbase, these will be the conorms of the lattice.

Consider the lattice L_0 specified by the following inner product matrix for a base v_1, v_2, v_3:

$$\begin{pmatrix} 3 & 1 & 1 \\ 1 & 4 & 2 \\ 1 & 2 & 5 \end{pmatrix}.$$

We adjoin v_0 so that $v_1 + v_2 + v_3 + v_0 = 0$; this augments the matrix so that the sum of each row and column is 0.

$$\begin{pmatrix} 3 & 1 & 1 & -5 \\ 1 & 4 & 2 & -7 \\ 1 & 2 & 5 & -8 \\ -5 & -7 & -8 & 20 \end{pmatrix}$$

Since some inner products are positive, this superbase is not obtuse. Equivalently, some p_{ij} are negative:

$$p_{01} = 5, \ p_{02} = 7, \ p_{03} = 8, \ p_{12} = -1, \ p_{13} = -1, \ p_{23} = -2.$$

Nevertheless, we shall write these "putative conorm numbers" on our "dual Fano plane", obtaining diagram (a).

Now our arguments in the previous section show that we can get to the corresponding numbers for an adjacent superbase by adding ϵ to the three numbers on a line containing $-\epsilon$ and 0, and subtracting ϵ from the four numbers off this line.

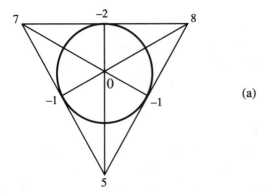

(a)

When we do this for the vertical line in diagram (a), we get diagram (b), which still has two negative numbers. So we repeat the process on the "circular" line of that figure, in order to obtain diagram (c). Two more such operations yield diagrams (d) and (e), and we have

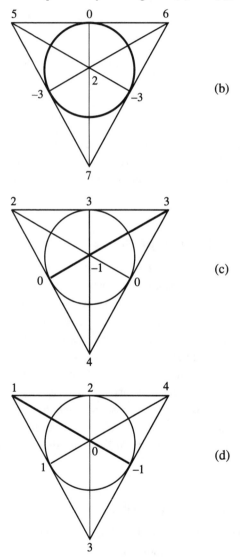

(b)

(c)

(d)

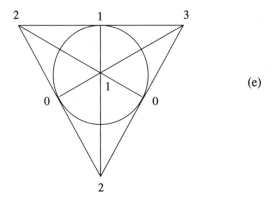

(e)

finished—since there is no negative number, this is the conorm diagram for an obtuse superbase. The algorithm always terminates because one of the "putative vonorms" is reduced at each stage.

Conorms and Selling parameters

Since a 3-dimensional lattice always has an obtuse superbase, it follows that one of the conorms is 0, while the others are the Selling parameters p_{ij} of some obtuse superbase. So in this dimension, our theory looks very much like the theory developed by Selling in 1874 and later generalized to higher dimensions by Voronoi and Delone. However, we have obtained a significant improvement by throwing in an extra 0: our definition of them shows that the seven conorms (unlike the six Selling parameters) vary continuously with the lattice. The conorms are also invariant in a sense in which the Selling parameters are not: two lattices described by conorms are equivalent just if their conorm functions are related by an isomorphism of the Fano plane.

The five shapes of Voronoi cell

In the 2-dimensional case, we found that the shape of the Voronoi cell was determined by the number of 0-conorms.

Three strictly positive conorms a, b, c yield a hexagonal Voronoi cell, which degenerates to a rectangle when we put $c = 0$. (It is impossible for two or more conorms to vanish.)

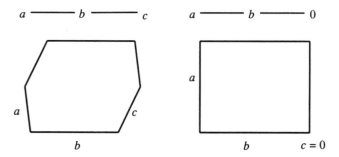

Similarly, in 3 dimensions, the various shapes of Voronoi cell correspond to the various possibilities for conorms of value 0, subject to the proviso that the support of the conorm function must not be contained in a line (for then the vonorm for that line would be 0). There are just five possibilities, which we shall now describe. When there are six nonzero conorms a, b, c, A, B, C, the Voronoi cell is a truncated octahedron, as shown below.

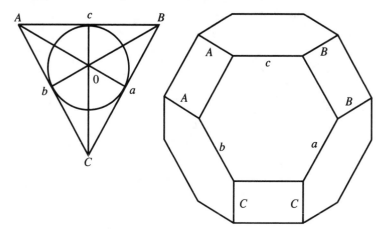

This has 6 classes of parallel edges, which correspond to the 6 conorms as also indicated in the figure.

When we put $c = 0$, all the corresponding edges (the horizontal ones in the figure) shrink to points, and we obtain a polyhedron which has been called an "elongated dodecahedron", but which we prefer to call the *hexarhombic dodecahedron*. It has four hexagonal and eight rhombic faces.

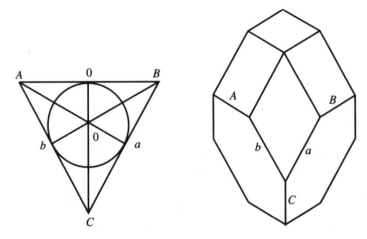

The third possibility is the *rhombic dodecahedron*, obtained by also putting $C = 0$ (shrinking the vertical edges). The three vanishing conorms now lie on a line in conorm space.

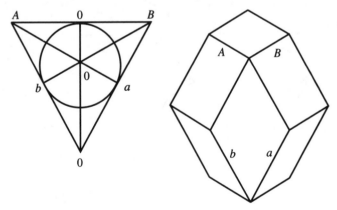

If instead we shrink the a edges of the hexarhombic dodecahedron, we obtain the fourth possibility, a hexagonal prism. It is characterized by having three vanishing conorms that do not lie on a line. The support of the conorm function consists of a line together with a further point not on that line.

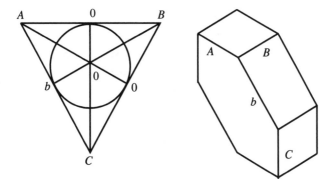

Finally, by shrinking either the a edges of the rhombic dodecahedron or the C edges of the hexagonal prism, we obtain the last possibility, the rectangular parallelepiped, or cuboidal box shape.

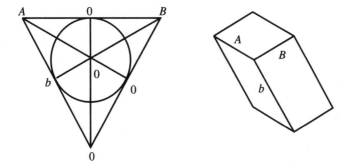

As any conorm vanishes, all the edges of a parallel class shrink to zero, since the formula for their squared length includes that conorm as a factor. At the same time, the other lengths and angles of the polyhedron change in a subtle way—for clarity, we have ignored this in the figures. Our names also ignore the fact that in the typical case, the "rhombs" will be more general parallelograms.

In summary: we have seen how to determine the Voronoi cell of a 2- or 3-dimensional lattice by its conorms. In the Afterthoughts, we give a brief outline of the the corresponding discussion in 4 dimensions.

The sphere-packing lattices

The lattices that have the smallest determinant for a given minimal norm are called the *sphere-packing lattices*, since they yield the densest lattice packings of equal spheres. If we take the minimal norm to be 2, the answers are known in up to 8 dimensions—they are the root lattices:

$$A_0, \ A_1, \ A_2, \ D_3, \ D_4, \ D_5, \ E_6, \ E_7, \ E_8.$$

The cases of dimension $n < 3$ are easy; 3 was settled by Gauss [Gau2] in 1831, 4 and 5 by Korkine and Zolotareff [KZ] in the 1870's, and 6, 7, 8 were later handled by Blichfeldt [Bli] before 1935. Blichfeldt's arguments were simplified by Mordell [Mor], who showed that 8 was easily deduced from 7, and by Vetčinkin [Vet]; but the calculations are still very complicated. (See [CSIII] for more information). We shall briefly sketch proofs for the cases $n = 2, 3$.

If a 2-dimensional lattice has three vonorms equal to 2 (the assumed minimal value), the shape of the lattice is determined, and it is A_2. However, if we only impose the condition that at most two vonorms be 2, there is still a variable parameter and it is easy to show that the determinant is decreased by reducing the third vonorm.

For $n = 3$, a similar argument shows that to achieve the minimal determinant, we must make at least six vonorms be 2, since there are six variable parameters in the matrix. But then the last one must be 4 (to make one of the conorms be 0), and again the lattice is determined and is D_3 (which is isomorphic to A_3).

Minkowski Reduction

Since we have described Voronoi reduction of 2 and 3-dimensional lattices in so much detail, it seems appropriate that we should at least mention the much more familiar notion of Minkowski reduction.

An integral base $\mathbf{e}_1, \ldots, \mathbf{e}_n$ is called a *Minkowski reduced* base for f just if each $f(\mathbf{e}_i)$ is the smallest value of $f(\mathbf{e}'_i)$ over all integral bases $\mathbf{e}_1, \ldots, \mathbf{e}_{i-1}, \mathbf{e}'_i, \mathbf{e}'_{i+1}, \ldots, \mathbf{e}'_n$. The positive definite form f is said to be *Minkowski reduced* if it is specified by its matrix with respect to a Minkowski reduced basis. This condition reduces to a finite set of matrix inequalities on the set of matrix entries. For the binary form $\begin{pmatrix} a & h \\ h & b \end{pmatrix}$, these are:

$$|2h| \leq a \leq b.$$

For the ternary form

$$\begin{pmatrix} a & h & g \\ h & b & f \\ g & f & c \end{pmatrix}$$

they are

$$a \leq b \leq c,$$

$$2|h| \leq a, \quad 2|g| \leq a, \quad 2|f| \leq b$$

and

$$2|h \pm g \pm f| \leq a + b.$$

As the dimension increases, the number of inequalities increases very rapidly, and they have only been written down as far as dimension 7.

The Little Methuselah Form

The Minkowski conditions are used in proving many number-theoretical theorems. As an example, we shall use them to prove an amusing little theorem:

Theorem. *The Little Methuselah Form*

$$F(x, y, z) = x^2 + 2y^2 + yz + 4z^2$$

represents every integer from 1 to 30, but fails to represent 31. Every integer-valued positive definite ternary form G not equivalent to this form F fails to represent some integer between 1 and 30.

Proof.   The topograph (below) for this form shows that

$$2, \ 4, \ 5, \ 7, \ 10, \ 14, \ 16, \ 19, \ 20, \ 25, \ 28, \ 32$$

are the smallest primitive values of $2y^2 + yz + 4z^2$, and multiplying these by squares we find the numbers on the top line to be *all* the values of the latter form up to 32.

```
0,  2,  4,  5,  7,  8,  10, 14, 16, 18, 19, 20, 25, 28, 32
1,  3,  5,  6,  8,  9,  11, 15, 17, 19, 20, 21, 26, 29
4,  6,  8,  9,  11, 12, 14, 18, 20, 22, 23, 24, 29, 32
9,  11, 13, 14, 16, 17, 19, 23, 25, 27, 28, 29,
16, 18, 20, 21, 23, 24, 26, 30, 32,
25, 27, 29, 30, 32,
```

The subsequent lines add squares to these, and show that indeed our Little Methuselah Form represents all integers from 0 to 32 except 31.

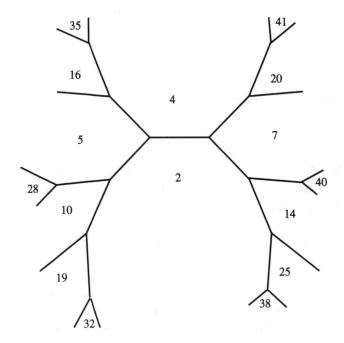

To show that no other form G can do so well, we suppose that the matrix for G is the generic 3×3 Minkowski reduced matrix discussed earlier. Since G represents 1, we see that a must be 1. Then, since G represents 2, we find that b is at most 2. Now we can list all possibilities for the binary subform $\begin{pmatrix} a & h \\ h & b \end{pmatrix}$ and we find that in every case there is some $m \leq 5$ not represented by this subform and so for the form G we must have $c \leq 5$. The Minkowski inequalities now leave only finitely many possibilities for G and we can check that each one not equivalent to F omits some number less than 31. $\quad\square$

This argument has not really explained just *why* every positive definite integer-valued ternary form *must* miss some integer. We shall show in the Postscript that, even over the rationals, a positive definite ternary form must fail to represent some integer.

Valediction

This lecture has generalized our previous discussion of binary positive definite quadratic forms to the 3-dimensional case. In four dimensions, the situation becomes considerably more complicated, as we describe in the Afterthoughts. Rather than having just one "primitive" (generic) shape of Voronoi cell, we have 3, only one of which has an obtuse superbase. The total number of shapes rises from 5 to 52. In 5 dimensions, there are 222 primitive shapes, and an extremely large (and still unknown) number of other shapes.

As regards the classification of integral positive definite quadratic forms: Minkowski reduction is very useful up to about 8 dimensions and we retain some degree of control up to about 24 dimensions using the gluing method described in the Afterthoughts to the Second Lecture. Beyond that, such forms seem to become inherently unclassifiable. The indefinite forms of rank at least 3, however, admit a complete classification by quite different methods, as we discuss in the next lecture.

Feeling the Form of a Four-Dimensional Lattice

Conorms and Selling parameters

In the body of the lecture, we showed that in 2 or 3 dimensions, the shape of the Voronoi cell was determined by the positions of the conorms of value 0. In 2 dimensions, the cell is rectangular or hexagonal, according as there is or is not a 0 conorm. In 3 dimensions, when the number of 0 conorms is

1	the cell is a truncated octahedron
2	the cell is a hexarhombic dodecahedron
3 (in line)	the cell is a rhombic dodecahedron
3 (not)	the cell is a hexagonal prism
4	the cell is a rectangular parallelipiped.

However, conorms really come into their own for 4-dimensional lattices. They enable us to give a simple description of the 52 types of lattices that were enumerated by Delone [Del], as corrected by Stogrin [Sto].

It is something of a coincidence that in the first few dimensions, the conorms are virtually equal to the Selling parameters. For a lattice of dimension

1, 2, 3, 4, 5, ..., n,..., there are

1, 3, 7, 15, 31, ..., $2^n - 1$,... conorms, but only

1, 3, 6, 10, 15, ..., $\frac{n(n+1)}{2}$,... Selling parameters.

In 1 and 2 dimensions, the conorms are precisely the Selling parameters, and in 3 dimensions, they are the Selling parameters supplemented by 0. A similar thing happens for *some* lattices in any number of dimensions: if a lattice has an obtuse superbase, its conorms are the Selling parameters supplemented by 0s.

Four-dimensional graphical lattices

Sixteen of the 4-dimensional cases are like this. They are parametrized by certain subgraphs of the complete graph K_5 on 5 vertices, in the following way. The vertices α, β, γ, δ, ϵ are five characters whose product is the trivial character, and we have

$$co(\alpha) = co(\beta) = co(\gamma) = co(\delta) = co(\epsilon) = 0$$

and $co(\lambda\mu) \neq 0$ just when there is an edge $\lambda\mu$ in the graph.

There is only one other case having any zero conorms. It corresponds to the complete bipartite graph $K_{3,3}$ on six nodes α, β, γ, and δ, ϵ, ζ. In this case $\alpha\beta\gamma$ and $\delta\epsilon\zeta$ are each the trivial character and we have precisely six zero conorms:

$$co(\alpha) = co(\beta) = co(\gamma) = co(\delta) = co(\epsilon) = co(\zeta) = 0$$

For any of these seventeen cases, we can indicate the exact shape of the lattice by marking the edges of the corresponding graph with the conorm values. We classify the seventeen "graphical" cases below, according to the number of independent parameters (one for each nonzero conorm). Representative graphs are shown on the next page.

parameters	lattices
10	K_5
9	$K_{3,3}$, $K_5 - 1$
8	$K_5 - 2$, $K_5 - 1 - 1$
7	$K_5 - 3$, $K_5 - 2 - 1$, $K_4 + 1$, C_{2221}
6	K_4, C_{321}, C_{222}, $C_3 + C_3$
5	C_5, $C_4 + 1$, $C_3 + 1 + 1$
4	$1 + 1 + 1 + 1$

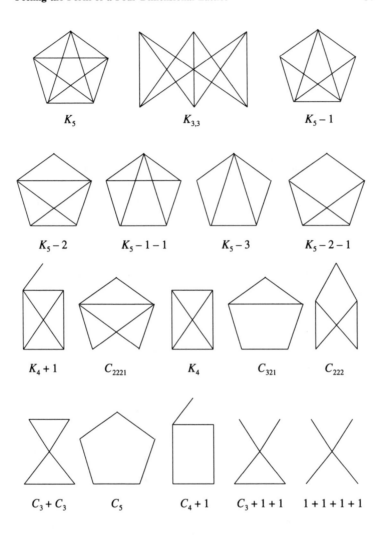

The remaining four-dimensional lattices

If no conorms are 0, then it turns out that precisely three of them are negative, and that these all have the same value, say

$$co(\alpha) = co(\beta) = co(\gamma) = -p,$$

and that $\alpha\beta\gamma$ is the trivial character. We can now display the (non-trivial) characters in an array

$$
\begin{array}{cccc}
 & \alpha & \beta & \gamma \\
\delta & \alpha\delta & \beta\delta & \gamma\delta \\
\epsilon & \alpha\epsilon & \beta\epsilon & \gamma\epsilon \\
\zeta & \alpha\zeta & \beta\zeta & \gamma\zeta
\end{array}
$$

in which we can suppose that $\delta\epsilon\zeta$ is the trivial character. By our convention, all three conorms in the top row will have values $-p$. It is a remarkable fact that the smallest conorm in each of the other three rows is p. The 35 cases are classified by the positions in which these minima occur. We say the lattice has type rst if the minimum occurs r, s, and t times in the three rows.

However, there might still be several inequivalent ways to position the conorms of value p: we display conorms for some lattices of type 322.

$$
\begin{bmatrix}
 & -p & -p & -p \\
p & p & p & \\
p & p & & \\
p & p & &
\end{bmatrix}
\qquad
\begin{bmatrix}
 & -p & -p & -p \\
p & p & & p \\
p & p & & \\
 & & p & p
\end{bmatrix}
$$

$$322+ \qquad\qquad 322-$$

$$
\begin{bmatrix}
 & -p & -p & -p \\
 & p & p & p \\
p & p & & \\
p & & p &
\end{bmatrix}
$$

$$322'$$

For any lattice of type 322, the characters of conorm p will be (say) $\delta_1, \delta_2, \delta_3$; ϵ_1, ϵ_2; ζ_1, ζ_2.

A digit 2 corresponds to a pair of characters in the same row, whose product will be some character from the top row. When two or more of rst are 2, then we add primes if necessary to indicate that the corresponding product characters are different. For instance, in our first two examples, $\epsilon_1\epsilon_2 = \zeta_1\zeta_2 = \alpha$ so no primes are needed. But in the third example, $\epsilon_1\epsilon_2 = \alpha$ and $\zeta_1\zeta_2 = \beta$, so we name it $322'$.

However, the top two cases are still combinatorially distinct: in the first case $(322+)$, the characters

$$\delta_1\epsilon_1\zeta_1, \ \delta_1\epsilon_2\zeta_2, \ \delta_2\epsilon_1\zeta_2, \ \delta_2\epsilon_2\zeta_1$$

are all trivial, but for the second case $(322-)$ there are only two trivial triples

$$\delta_3\epsilon_1\zeta_2, \ \delta_3\epsilon_2\zeta_1.$$

In general, we append $+$ or $-$ if $\delta_i\epsilon_j\zeta_k$ is the trivial character for more or less than the "expected" number $pqr/4$ of triples ijk.

We now list the 35 non-graphical cases according to the number of independent parameters:

parameters	lattices
10	$111+$, $111-$
9	$211+$, $211-$
8	$311+$, $311-$, $221+$, $221-$, $22'1$
7	411, $321+$, $321-$, $222+$, $222-$, $222'$, $22'2''$
6	421, $331+$, $331-$, $322+$, $322-$, $322'$
5	431, 422, $422'$, $332+$, $332-$
4	441, 432, $333+$, $333-$
3	442, 433
2	443
1	444

We see that just the three cases $(K_5, 111+, 111-)$ have the full number 10 of parameters. Voronoi called such generic cases *primitive*. In 5 dimensions, there are already 222 primitive types, found by Baranovskii and Ryshkov in 1973 (as later corrected by Engel). The Voronoi cell of a primitive lattice has the maximal number $2(2^n - 1)$ of faces. However, in 4 dimensions, type $K_{3,3}$ also has 30 faces, and in 5 dimensions, there are 3 nonprimitive lattices whose Voronoi cells still have 62 faces.

The Primary Fragrances

Introduction

In the Third Lecture, we classified definite forms in up to three dimensions by a process which is equivalent to examining their Voronoi cells geometrically. But the essential essences of a quadratic form are arithmetical! By considering congruences modulo powers of primes, it is possible to write down arithmetical invariants that tell us a lot about the form. More precisely, they completely solve the problem of rational equivalence for all forms.

There is a way to enlarge the field Q of rationals to certain larger fields Q_p of "p-adic rationals", one for each prime p. Although the p-adics are the standard basis for the theory, we don't actually use them in the Lecture. The reader who wants a deeper understanding will find a discussion in the Afterthoughts.

A quadratic form over Q is rather like a bouquet of flowers, each flower being the corresponding form over one of the fields Q_p. From the fragrances of these flowers we can recover the structure of the rational form.

In the first half of this Lecture, we shall give the complete theory for rational equivalence of quadratic forms. The second and more detailed half is about the integral invariants. It gives a new invariance proof using the concept of audibility from the Second Lecture.

Equivalence over Q; diagonalization

If we allow transformations with rational coefficients, for example, that of replacing the basis vector e_2 by $e_2 - \frac{3}{2}e_1$, the equivalence problem becomes easier. For example, let

$$
\begin{array}{c c}
 & \begin{array}{cccc} e_1 & e_2 & e_3 & e_4 \end{array} \\
\begin{array}{c} e_1 \\ e_2 \\ e_3 \\ e_4 \end{array} &
\left(\begin{array}{cccc}
2 & 3 & 4 & 5 \\
3 & 6 & 7 & 8 \\
4 & 7 & 9 & 10 \\
5 & 8 & 10 & 11
\end{array}\right)
\end{array}
$$

represent a 4-dimensional quadratic form, which we want to diagonalize. It is best to think of the matrix as representing the inner products of four vectors e_1, e_2, e_3, e_4 as shown. What we want is to find a base of four mutually orthogonal vectors.

To this end, we consider first an intermediate base consisting of $e_1' = e_1$ and three vectors

$$
e_2' = e_2 - \tfrac{3}{2}e_1, \quad e_3' = e_3 - 2e_1, \quad e_4' = e_4 - \tfrac{5}{2}e_1
$$

orthogonal to e_1. The inner products of these with the original four are given by the matrix

$$
\begin{array}{c c}
 & \begin{array}{cccc} e_1 & e_2 & e_3 & e_4 \end{array} \\
\begin{array}{c} e_1' \\ e_2' \\ e_3' \\ e_4' \end{array} &
\left(\begin{array}{cccc}
2 & 3 & 4 & 5 \\
0 & \frac{3}{2} & 1 & \frac{1}{2} \\
0 & 1 & 1 & 0 \\
0 & \frac{1}{2} & 0 & -\frac{3}{2}
\end{array}\right)
\end{array}
$$

obtained from the initial matrix by subtracting $\frac{3}{2}$ times the first row from the second, twice the first row from the third, and $\frac{5}{2}$ times it from the fourth. To get the inner products of the new vectors against each other, we must perform the corresponding column operations (which just clear three matrix entries):

$$
\begin{array}{c} \\ \mathbf{e}_1' \\ \mathbf{e}_2' \\ \mathbf{e}_3' \\ \mathbf{e}_4' \end{array}
\begin{array}{cccc} \mathbf{e}_1' & \mathbf{e}_2' & \mathbf{e}_3' & \mathbf{e}_4' \\ \left(\begin{array}{cccc} 2 & 0 & 0 & 0 \\ 0 & \frac{3}{2} & 1 & \frac{1}{2} \\ 0 & 1 & 1 & 0 \\ 0 & \frac{1}{2} & 0 & -\frac{3}{2} \end{array}\right) \end{array},
$$

and we have finished one stage of the diagonalization.

Since the next diagonal entry $\frac{3}{2}$ is nonzero, we can continue in a similar way, subtracting multiples of the second row from the third and fourth rows, and then performing the corresponding column operations so as to obtain the matrix

$$
\begin{pmatrix} 2 & 0 & 0 & 0 \\ 0 & \frac{3}{2} & 0 & 0 \\ 0 & 0 & \frac{1}{3} & -\frac{1}{3} \\ 0 & 0 & -\frac{1}{3} & -\frac{5}{3} \end{pmatrix},
$$

which after just one more stage becomes diagonal:

$$
\begin{pmatrix} 2 & 0 & 0 & 0 \\ 0 & \frac{3}{2} & 0 & 0 \\ 0 & 0 & \frac{1}{3} & 0 \\ 0 & 0 & 0 & -2 \end{pmatrix}.
$$

Each of the above steps post-multiplies the matrix of the form by some rational matrix, and then pre-multiplies by the transpose of that matrix. We have therefore proved that there is some matrix M for which

$$
M^t \begin{pmatrix} 2 & 3 & 4 & 5 \\ 3 & 6 & 7 & 8 \\ 4 & 7 & 9 & 10 \\ 5 & 8 & 10 & 11 \end{pmatrix} M = \begin{pmatrix} 2 & 0 & 0 & 0 \\ 0 & \frac{3}{2} & 0 & 0 \\ 0 & 0 & \frac{1}{3} & 0 \\ 0 & 0 & 0 & -2 \end{pmatrix}.
$$

In other words, our form is rationally equivalent to the form $\mathrm{diag}[2, \frac{3}{2}, \frac{1}{3}, -2]$, which we shorten to $[2, \frac{3}{2}, \frac{1}{3}, -2]$.

The same approach works for any form with some nonzero diagonal entry. If all diagonal terms are 0 but some off-diagonal term, say a_{ij}, is nonzero, then we modify the matrix by what we call the *pushover trick*; add the jth row to the ith and then the jth column to the ith—this produces a nonzero diagonal entry $2a_{ij}$. If *all* entries are 0, then the form already *is* diagonal. We have therefore shown that over the rationals, any quadratic form can be diagonalized.

The invariant problem

But if in our example we started from the lower right hand entry and eliminated the bottom row and right column, we'd get

$$\begin{pmatrix} -\frac{3}{11} & -\frac{7}{11} & -\frac{6}{11} & 0 \\ -\frac{7}{11} & \frac{2}{11} & -\frac{3}{11} & 0 \\ -\frac{6}{11} & -\frac{3}{11} & -\frac{1}{11} & 0 \\ 0 & 0 & 0 & 11 \end{pmatrix},$$

and then, continuing from the bottom up, we'd eventually obtain

$$\begin{pmatrix} 2 & 0 & 0 & 0 \\ 0 & 1 & 0 & 0 \\ 0 & 0 & -\frac{1}{11} & 0 \\ 0 & 0 & 0 & 11 \end{pmatrix}.$$

So we have proved that

$$[2, \tfrac{3}{2}, \tfrac{1}{3}, -2] = [2, 1, -\tfrac{1}{11}, 11],$$

where $=$ denotes rational equivalence.

We need some way of telling just when two diagonal forms are equivalent over the rationals! This is not at all obvious: $2x^2 + 2y^2$ equals $(x + y)^2 + (x - y)^2$ and so is equivalent to $x^2 + y^2$, which is also equivalent to $5x^2 + 5y^2$, but not to $3x^2 + 3y^2$!

We shall settle such questions in terms of certain invariants called the *p-adic signatures:*

$$\sigma_{-1}, \ \sigma_2, \ \sigma_3, \ \sigma_5, \ \sigma_7, \ \dots$$

together with the determinant of the form. Note that the determinant itself is *not* quite an invariant; but it *is* invariant up to multiplication by nonzero rational squares.

The signature σ_p determines what we called in the introduction the "fragrance" of the form over $\mathbf{Q}_p$. This is appropriate, since in fact the word "signature" is a term used in the scent industry!

The signatures of a quadratic form

Every nonzero rational number can be factored uniquely as a product of prime powers. For example,

$$-4\tfrac{1}{2} = (-1) \times 2^{-1} \times 3^2.$$

Note that we treat -1 as a prime just like all the others, but it only has two powers 1 and -1, so it's a prime of order 2.

When we factor a number into prime powers, the power of p is called the p-part of the number. We will define the p-signatures in terms of these p-parts. For instance, the ordinary signature introduced by Sylvester is the sum of all the -1-parts of the diagonal entries. We shall call it σ_{-1}, and will show how to define analogous signatures, σ_p, for the other primes. Since we can multiply the diagonal entries by nonzero squares, we shall suppose them to be integers.

Let then f be a diagonal form with integer entries. we define $\sigma_p(f)$ for an odd prime p by adding up the p-parts of the entries, and then adding 4 for each p-adic antisquare. A number $a = p^r a'$ is a p-adic antisquare just if r is odd and a' is not a quadratic residue mod p (so a fails to be a square in both ways; its p-part isn't a square, and what's left isn't even a square modulo p). For $p \neq -1$, these signatures are only defined modulo 8.

So for instance, the 3-signature of $[10, -9, 21, 6]$ is

$$1 + 9 + 3 + 3$$
$$+ 4$$

which is 4 (mod 8), since 6 is a 3-adic antisquare. Also, its 5-signature is

$$5 + 1 + 1 + 1$$
$$+4$$

again 4 (mod 8). However, its 7-signature is

$$1 + 1 + 7 + 1$$

$$+ 4$$

namely 6 (mod 8).

The 2-*signature*, or *oddity*, is obtained instead by adding up the **odd parts** (including their signs) of the entries (things are always upside down for $p = 2$), and then adding 4 for each entry of form $2^{odd}(8k \pm 3)$—these are the 2-*adic antisquares*.

For instance, the 2-signature of the above form is

$$5 - 9 + 21 + 3$$

$$+4 \qquad\qquad + 4$$

which is 4 (mod 8).

Finally, the -1-signature for this form is of course

$$1 - 1 + 1 + 1,$$

namely the integer 2. It is important to realize that although the -1-signature is an absolute integer, the others are determined only modulo 8.

The Hasse-Minkowski theorem and the global relation

These signatures completely solve the rational equivalence problem. *Two forms f and g of the same dimension are rationally equivalent if and only if their determinants are equal (modulo squares) and they have the same p-signatures for all primes $-1, 2, 3, 5, \ldots$.* This is the celebrated Hasse-Minkowski theorem, recast in a somewhat unorthodox manner.

Let us use this theorem to check the equivalence we found between

$$[2, \frac{3}{2}, \frac{1}{3}, 2] \qquad \text{and} \qquad [2, 1, -\frac{1}{11}, 11].$$

Multiplying the entries by squares, we obtain the equivalent forms

$$[2, \quad 6, \quad 3, \quad -2] \text{ and } [2, \quad 1, \quad -11, \quad 11],$$

and compute the signatures:

$$\sigma_2: \quad \begin{matrix} 1 & +3 & +3 & -1 \\ & +4 & & \end{matrix} \equiv 2 \equiv 1 \quad +1 \quad -11 \quad +11$$

$$\sigma_{-1}: \quad 1 \quad +1 \quad +1 \quad -1 = 2 = 1 \quad +1 \quad -1 \quad +1$$

$$\sigma_3: \quad \begin{matrix} 1 & +3 & +3 & +1 \\ & +4 & & \end{matrix} \equiv 4 \equiv 1 \quad +1 \quad +1 \quad +1$$

$$\sigma_{11}: \quad 1 \quad +1 \quad +1 \quad +1 \equiv 4 \equiv \begin{matrix} 1 & +1 & +11 & +11 \\ & & +4 & \end{matrix}$$

and

$$\sigma_p: \quad 1 \quad +1 \quad +1 \quad +1 \equiv 4 \equiv 1 \quad +1 \quad +1 \quad +1$$

for every other p, all congruences being modulo 8.

Our "p-adic signature" is in fact an invariant under transformations over the larger field $\mathbf{Q}_p$ of p-adic rationals, so the theorem implies the usual statement:

Theorem. *Two forms f and g are equivalent over $\mathbf{Q}$ if and only if they are equivalent over $\mathbf{Q}_p$ for all primes p (including -1).*

The p-adic structures for different p are nearly, but not quite, independent. It turns out that there is just one relation between them, which we shall call *The Global Relation* (it usually arises in the form of a "product formula"). To define this, we shall introduce some slightly modified invariants, the *p-excesses* $e_p(f)$, defined by

$$e_p(f) = \sigma_p(f) - \dim(f), \qquad \text{if } p \neq 2$$
$$e_2(f) = \dim(f) - \sigma_2(f).$$

Then the global relation asserts that

Theorem. *The sum of the p-excesses for all primes p (including -1) is a multiple of 8.*

This is essentially the only relation between the determinant and the p-excesses. For suppose that for each individual p there exists a form with the required values of d and the p-excesses. Then we shall also prove in the Postscript that there will be a single rational form that achieves all these values simultaneously precisely when the global relation holds.

In the next few sections, we shall prove the Hasse-Minkowski theorem by a sequence of reductions.

Reduction to the case of trivial invariants

We next show that we need only consider forms with *trivial invariants,* that is to say, the determinant is a square, and all dimensions and p-signatures are multiples of 8.

In the next few sections, we shall prove that a form with trivial invariants is equivalent to some form of shape

$$[\pm 1, \pm 1, \ldots, \pm 1].$$

The reduction uses a result called *Witt's cancellation law,* which says that, for a nonsingular h, if $h \oplus f$ is equivalent to $h \oplus g$ over a field not of characteristic 2, then f is equivalent to g. If f and g have the same invariants, then the forms

$$g \oplus g \oplus g \oplus g \oplus g \oplus g \oplus g \oplus f$$

and

$$g \oplus g \oplus g \oplus g \oplus g \oplus g \oplus g \oplus g$$

will have (the same) trivial invariants. So the above statement proves that these two forms are equivalent, and Witt's cancellation law will then tell us that f is equivalent to g.

Proving Witt cancellation

In proving Witt's cancellation law, it suffices to take the case of a 1-dimensional form h; in other words, to show that if

$$[a, b, c, \ldots] = [a, b', c', \ldots],$$

and $a \neq 0$, then

$$[b, c, \ldots] = [b', c', \ldots].$$

The first equality may be expressed by saying that the space generated by orthogonal vectors $e_1, e_2, e_3, \ldots$ with

$$e_1 \cdot e_1 = a, \qquad e_2 \cdot e_2 = b, \qquad e_3 \cdot e_3 = c \quad \ldots,$$

has another base of orthogonal vectors $e'_1, e'_2, \ldots$ with

$$e'_1 \cdot e'_1 = a, \qquad e'_2 \cdot e'_2 = b', \qquad e'_3 \cdot e'_3 = c', \quad \ldots$$

It will suffice to find a symmetry of this space that takes e_1 to e'_1, for this symmetry will necessarily take the space $< e_2, e_3, \ldots >$ orthogonal to e_1 onto the space $< e'_2, e'_3, \ldots >$ orthogonal to e'_1. Now the reflection in the hyperplane perpendicular to a vector r is given by the formula

$$x \mapsto x - \frac{2(x, r)}{(r, r)} r.$$

It is easily verified that this does preserve our quadratic form (x, x). If $r = e_1 - e'_1$, then this reflection does indeed take e_1 to e'_1. However, there is a problem. We can only reflect in a vector of nonzero length, but the form we are working with may be indefinite, so it might happen that the norms of some nonzero vectors are 0. But if r has norm 0, then we can reflect in the hyperplane perpendicular to the vector $s = e_1 + e'_1$, which takes e_1 to $-e'_1$, and then negate. Not both (r, r) and (s, s) can be 0, since we have, by the Apollonian identity,

$$(r, r) + (s, s) = 2(e_1, e_1) + 2(e'_1, e'_1) = 4a \neq 0.$$

Replacing p-terms

So now we can suppose that we have a form f with trivial invariants and want to make all of the entries ± 1. In fact we shall do this instead for a suitable direct sum

$$f \oplus [\pm 1, \pm 1, \ldots, \pm 1]$$

and then use Witt cancellation to deduce it for f. We take f in diagonal form with squarefree entries, and let p be the largest prime that occurs in any entry. The *p-terms* are those diagonal entries of the form

$$pq_0 q_1 q_2 \cdots q_t \qquad (-1 \leq q_i < p).$$

The first order of business will be to simplify such a p-term by "replacing" its cofactor $q_0 q_1 q_2 \cdots q_t$ by something more manageable.

The Replacement Lemma.   *Let $p \geq 3$. Then given a p-term*

$$pqq_1 q_2 q_3 \ldots = pqt, \qquad (-1 \leq q < p)$$

and a number q' with

$$-1 \leq q < p, \quad -1 \leq q' < p,$$

qq' *congruent to a square mod* p,

we can replace the terms pqt, 1, -1 by $pq't$, $*$, $*$ where the prime factors of the two starred terms are all less than p.

Proof.   There is an integer x with

$$qq' \equiv x^2 \pmod{p}$$

and $|x| < \frac{1}{2}p$. Then we can write $qq' = x^2 - py$, wherein $-p^2 < py < p^2$, so that $|y| < p$. Then

$$pqt(\frac{q'}{x})^2 + yq't(\frac{p}{x})^2 = pq't.$$

Hence the form $[pqt, yq't]$ represents $pq't$, so that

$$[pqt, yq't] = [pq't, *],$$

where the number $*$ must be yqt times a perfect square, by considering the determinant. Also, the form $[1, -1]$, namely $x^2 - y^2$, represents every number, so that

$$[-1, 1] = [m, -m]$$

for any $m \neq 0$ by the same argument.

It follows that the form

$$[pqt, 1, -1]$$

is rationally equivalent to

$$[pqt, yq't, -yq't]$$

and this in turn is equivalent to

$$[pq't, yqt, -yq't],$$

as required.   □

The coup de grâce

Now let $u = r + 1$ be the least quadratic nonresidue mod p. Then given a p-term $pq_0q_1q_2 \ldots q_s$, we can successively replace each q_i by either 1 or u, and so replace the entire p-term by p or pu, since squares may be cancelled.

Now since $[p, pr]$ represents the number pu, it is equivalent to $[pu, pru]$. But $r = u - 1$ is a quadratic residue (by definition of u), and hence can be replaced by 1. Thus, by successive replacements, we can change a pair of p-terms p, p through p, pr and pu, pru to pu, pu.

In short, we can "flip" a couple of p-terms p, p into pu, pu or vice-versa. This proves

The Freedom Lemma.   *Of any two p-terms, the first can be chosen freely.*

For if they are the same, we can flip them, and if they are different, we can swap them!   □

Now there is at least one p-term by hypothesis, and in fact at least two, since the determinant is a square. The first can be whatever we like: make it $-p$. If there are three or more, make the second one be p, and then eliminate two of them using $[p, -p] = [1, -1]$.

In the remaining case when there are exactly two p-terms, they must either be $-p$, p, when we can eliminate them, or $-p$, pu. But the p-signatures of these two cases clearly differ by 4, and since the p-signature must be trivial for the case with $-p$, p, the second case cannot be what we face.

We summarize: if the largest prime p that occurs is at least 3, then we can eliminate all the p-terms. So eventually we can reduce p to 2, where the argument is even easier.

Namely, we can eliminate all the 2-terms (of which there are evenly many) using the equivalences

$$[2, 2] = [1, 1],$$
$$[2, -2] = [1, -1],$$
$$[-2, -2] = [-1, -1].$$

Eventually, we reduce to $p = -1$ and have put f into the required form

$$[\pm 1, \pm 1, \ldots, \pm 1].$$

If now f and g have the same invariants (including σ_{-1}), then

$$g \oplus g \oplus g \oplus g \oplus g \oplus g \oplus g \oplus f$$

and

$$g \oplus g \oplus g \oplus g \oplus g \oplus g \oplus g \oplus g$$

have the same trivial invariants, so can both be reduced to this kind of form. In fact they become the *same* form since they have the same rank and signature. We can now use Witt cancellation to deduce that f is equivalent to g.

Other versions of the Hasse-Minkowski invariant

In place of our signatures, most authors use invariants that take just the two values 1 and -1, but there is a bewildering variety of particular such invariants in the literature. Four cases seem to deserve names, namely the *Minkowski unit*

$$C_p(f) = \left(\frac{b_1}{p}\right)\left(\frac{-b_2}{p}\right)\left(\frac{b_3}{p}\right)\left(\frac{-b_4}{p}\right)\ldots,$$

the *conjugate Minkowski unit*

$$C_p(-f) = \left(\frac{-b_1}{p}\right)\left(\frac{b_2}{p}\right)\left(\frac{-b_3}{p}\right)\left(\frac{b_4}{p}\right)\cdots,$$

the *exclusive Hilbert product*

$$(f)_p = \prod (a_i, a_j)_p \qquad (i < j),$$

and the *inclusive Hilbert product*

$$(f)_p^* = \prod (a_i, a_j)_p \qquad (i \le j).$$

Here the form f is $[a_1, a_2, a_3, \ldots]$, and b_1, $b_2, \ldots$ are the p'-parts of those a_j that are divisible by *odd* powers of p. The *Legendre symbol* $\left(\frac{b}{p}\right)$ is 1 or -1 according as b is or is not a square modulo p, and the *Hilbert norm residue symbol* $(a_i, a_j)_p$ is defined to be 1 or -1 according as the form $[a_i, a_j]$ has or has not the same p-signature as $[1, a_i a_j]$.

The following remarks may help the reader to understand the situation. For any dimension $n > 2$ and determinant d, there are just two inequivalent forms over $\mathbf{Z}_p$: the one you first thought of, and the other one.

Now you may define your own personal invariant, by saying that it takes the value $+1$ for the form you first thought of, and -1 for the other one. If the form you first thought of happened to be one of those below, then your invariant is the appropriate one of the above four:

$$
\begin{array}{lll}
[d_p, d_{p'}, 1, 1, 1, \ldots, 1] & \text{for} & C_p(f) \\
[-d_p, -d_{p'}, 1, 1, \ldots, 1] & \text{for} & C_p(-f) \\
[d, 1, 1, \ldots, 1] & \text{for} & (f)_p \\
[-d, -1, 1, \ldots, 1] & \text{for} & (f)_p^*.
\end{array}
$$

Invariants for integral forms

This concludes for the moment our discussion of rational forms. We have not yet proved that our p-signatures really *are* invariants—we shall talk about this in the Afterthoughts to this lecture. Nor have we discussed the existence of forms with prescribed invariants, or the relations satisfied by the invariants—we shall return to this in the Postscript. In the remaining sections of this lecture, we shall produce some invariants for quadratic forms over the integers.

The reader who prefers to go straight to the Postscript may do so, since it makes no use of the theory to which we now proceed.

p-adic diagonalization and p-adic symbols

We have shown how to diagonalize any form over the rationals. It turns out that for any prime $p \geq 3$, we can perform this diagonalization without ever needing to divide by p. This is called *p-adically integral diagonalization*. [In the Afterthoughts, we shall see that it corresponds to diagonalization over the ring $\mathbf{Z}_p$ of *p-adic integers*.]

For $p = -1$, the corresponding notion is that of diagonalization over the ring $\mathbf{R}$ of real numbers. [In this context, $\mathbf{R}$ is also called $\mathbf{Z}_{-1}$ or $\mathbf{Q}_{-1}$. This is well-known linear algebra, and Sylvester's Law of Inertia asserts that the numbers n_+ and n_- of positive and negative diagonal terms are invariants. We shall describe this situation by saying that the -1-*adic symbol* is $(+)^{n_+}(-)^{n_-}$.]

For $p \geq 3$, we proceed as follows. In the normal case, when the power of p that divides some diagonal entry is the smallest power of p that divides *any* entry, we can start the diagonalization using that entry, since this will not require division by p. If the smallest power of p arises only in some off-diagonal entries, for example in a_{ij}, we can return to the normal case by the pushover trick of adding the jth column to the ith column and the jth row to the ith row.

We conclude that for any odd p, we can p-adically diagonalize forms to the shape

$$[a, b, c, \ldots, pa', pb', \ldots, p^2 a'', \ldots, \ldots],$$

say, where all the numbers $a, b, c, \ldots, a', b', \ldots, a'', \ldots$ are prime to p. (This includes the case $p = -1$ if by "prime to -1" we mean "positive".)

The forms

$$f_1 = [a, b, c, \ldots],$$
$$f_p = [a', b', c', \ldots],$$
$$f_{p^2} = [a'', b'', c'', \ldots],$$

$$\ldots$$

are called the *Jordan constituents* of f, and the expression

$$f_1 \oplus p f_p \oplus \cdots \oplus q f_q \oplus \cdots$$

is its *Jordan decomposition*. The Jordan constituents are p-adic *unit forms*, that is to say, their determinant is not divisible by p (or for $p = -1$, they are positive definite).

It turns out that for $p \neq 2$, the dimensions, determinants (up to p-adic squares as usual), and p-signatures of the Jordan constituents are a complete set of invariants for p-adic equivalence. The *p-adic symbol* is a nice way to encapsulate all of this information. The typical p-adic symbol looks like

$$1^{\pm l} p^{\pm k} \cdots q^{\pm n} \cdots$$

In general, we have a formal product of factors $q^{\pm n}$ where q is a power of p, n is the dimension of f_q, and the sign is the Legendre symbol $\left(\frac{\det(f_q)}{p}\right)$.

For example, $[1, 2, 3]$ has 3-adic symbol $1^{-1}3^{+1}$ since 2 is a quadratic non-residue modulo 3.

2-adic Jordan decomposition; the 2-adic symbol

For $p = 2$ things can get more complicated. If there *is* a diagonal entry in the matrix that is divisible by the smallest power of 2 in any entry, then we can proceed as before. However, we cannot reduce to this case when the smallest power of 2 occurs only in some off-diagonal entries a_{ij}, because the pushover trick involves $2a_{ij}$ rather than a_{ij}.

We must just accept the situation in such cases, when we can suppose that the leading 2×2 matrix has the form

$$q \begin{pmatrix} a & h \\ h & b \end{pmatrix} \qquad a, \ b \text{ even}, \ h \text{ odd}$$

where q is the smallest power of 2 dividing any entry.

Instead of diagonalizing f, we find ourselves expressing it as a direct sum of diagonal forms and 2×2 matrices of the above shape, for varying powers of q.

However, since $ab - h^2$ here is odd, we still obtain a Jordan decomposition

$$f_1 \oplus 2f_2 \oplus 4f_4 \oplus \cdots \oplus qf_q \oplus \cdots$$

in which each f_q is a 2-adic unit form. We can further suppose either that f_q has all its diagonal entries even (this is the *Type II* case) or (the *Type I* case) is of positive dimension and has an odd diagonal entry (when it can be diagonalized with all diagonal entries odd). The *2-adic symbol* encapsulates all of this information. It is a formal product of terms

$$q_t^{\pm n}$$

where the sign is the Jacobi symbol $\left(\frac{2}{\det(f_q)} \right)$, n is the dimension, and $t = \infty$ for a Type II form f_q, and $t = \text{trace}(f_q)$ modulo 8 when f_q is Type I and diagonal.

It is helpful to note that

The sign is $+$ when $\det(f_q) \equiv \pm 1 \pmod 8$,

The sign is $-$ when $\det(f_q) \equiv \pm 3 \pmod 8$.

For example, if

$$f = [1, 3, 5] \oplus 2 \begin{pmatrix} 2 & 1 \\ 1 & 2 \end{pmatrix} \oplus 4 \, [1, 1]$$

then its 2-adic symbol is

$$1_1^{+3} 2_\infty^{-2} 4_2^{+2}$$

since

$$1 \times 3 \times 5 \equiv \pm 1$$
$$1 + 3 + 5 \equiv 1$$, $$\det \begin{pmatrix} 2 & 1 \\ 1 & 2 \end{pmatrix} \equiv \pm 3,$$ $$1 \times 1 \equiv \pm 1$$
$$1 + 1 \equiv 2$$,

where all congruences are modulo 8.

The numbers mentioned are invariants of the Jordan constituents f_q. However, there is a difficulty. In the case $p = 2$, a form may have another Jordan decomposition $f_1' \oplus 2f_2' \oplus \cdots$, and the new Jordan constituents f_q' may *not* have the same invariants as the f_q. There are two rules for modifying 2-adic symbols to cope with this.

The *oddity-fusion rule* says that for any contiguous chain of forms $f_q, f_{2q}, f_{4q}, \ldots$, with finite t's, we can change the t's of the individual forms $f_q, f_{2q}, f_{4q}, \ldots$ in any way that leaves their sum invariant.

The *sign-walking rule* says that we can change the signs of two forms f_q and $f_{2^r q}$, provided that no two consecutive forms among

$$f_q, \quad f_{2q}, \quad \ldots, \quad f_{2^{r-1}q}, \quad f_{2^r q}$$

are both of type II, at the cost of changing some of the t's.

We "walk" from f_q to $f_{2^r q}$; every time we have to make a step from f_Q to f_{2Q}, at least one of these forms will be type I, and we add 4 to t for the odd form. (We cannot change the number $t = \infty$ for an even form.)

It is a fact that two forms are 2-adically integrally equivalent just if their 2-adic symbols are equivalent up to these modifications.

The genus

Two integral quadratic forms are said to be in the same *genus* just if they are p-adically equivalent for each p. Since our p-adic symbol is a complete invariant for p-adic equivalence, we may summarize our assertions in the

Theorem. *Two quadratic forms f and g are in the same genus if and only if they have the same p-adic symbols for each p, including -1.*

(Of course, "the same 2-adic symbol" means "2-adic symbols related to each other by oddity-fusion and sign-walking".)

The arguments in later sections show how the invariance of our symbols is proved.

p-adic Gauss means

In the Second Lecture, we defined some invariants called Gauss means. In later work, it turns out to be very convenient to redefine these slightly, writing $GM(f)$ for what we would formerly call "the Gauss mean of $f/2$", namely the mean of all the numbers

$$e^{\pi i N(\mathbf{v})} \qquad (\mathbf{v} \in L^*),$$

where L^* is the lattice dual to L. Of course, in this lecture it is also appropriate to pick out the parts of this that are invariant under p-adic integral transformations for each p, so we shall define also the p-*adic Gauss mean* $GM_p(f)$ to be the mean of the p-parts of the same numbers

$$e^{\pi i N(\mathbf{v})} \qquad (\mathbf{v} \in L^*).$$

[Note that any root of unity α factorizes uniquely as the product $\beta\gamma$ of two others, where the order of β is a power of p and that of γ is prime to p. The number β is the p-*part* of α, and γ is its p'-*part*.]

The p-adic Gauss means really do depend only on the equivalence class of f over the p-adic integers $\mathbf{Z}_p$. However, as we pointed out at the start of this lecture, we do not actually need the p-adic numbers. So we shall only explain briefly why they are invariant under rational p-adic integral transformations—that is, rational transformations that do not involve division by p. The reason is that such transformations multiply the terms in the Gauss mean only by roots of unity whose orders are prime to p, and so they do not change the p-parts at all.

Now we shall show that for $p \geq 3$ the p-excess can be computed from the p-adic Gauss mean, and in particular therefore that it really is a p-adic invariant.

The Gauss mean that we computed for the 1-dimensional form $[p]$ in the Second Lecture had value 1 or i divided by $\sqrt{p}$. But we have now changed the definition by removing a factor of 2 in the exponents—what difference does this make? The answer for Gauss sums is well-known—if we introduce a factor of k (prime to p) in the exponents, the Gauss sum is multiplied by the Legendre symbol $\left(\frac{k}{p}\right)$, We can therefore easily compute the new Gauss mean for the form $[p]$: according as

p is congruent to	1	or	3	or	5	or	7	$\pmod 8$
old Gauss mean is	1	or	i	or	1	or	i	$\div\sqrt{p}$,
symbol $\left(\frac{2}{p}\right)$ equals	1	or	-1	or	-1	or	1,	and so
new Gauss mean is	1	or	$-i$	or	-1	or	i	$\div\sqrt{p}$.

We can summarize this in a simple rule:

The p-adic Gauss mean for the form $[p]$ is

$$\frac{1}{\zeta^{p-1}\sqrt{p}},$$

where ζ is the eighth root of unity $(1+i)/\sqrt{2}$, so $\zeta^2 = i$.

The p-adic Gauss mean for the modified form $[kp]$ (where k is prime to p) is $\left(\frac{k}{p}\right)$ times this, which we can write as $1/(\zeta^m\sqrt{p})$, where m is

$$p-1 \qquad \text{or} \qquad p-1+4,$$

according as

$$k \text{ is} \qquad \text{or} \qquad \text{is not}$$

a square modulo p.

But the number displayed here is just the p-excess of $[kp]$! The similar calculations for forms $[kp^n]$ establish more generally that

Theorem. *For $p \geq 3$, the p-adic Gauss mean for a diagonal form*

$$f = [a_1, a_2, \ldots, pb_1, pb_2, \ldots, p^2c_1, \ldots]$$

is

$$\frac{1}{\zeta^{e_p(f)}\sqrt{\det_p(f)}},$$

where $\det_p(f)$ is the p-part of $\det(f)$.

Since the p-adic Gauss mean is an invariant of p-adic integral equivalence, this also gives an invariant definition of $e_p(f)$, and so proves that $e_p(f)$ is an invariant of f; indeed an invariant under (rational) p-adically integral transformations.

Audibility of the p-adic symbol

We can extract still more information by considering the forms f/p, $f/p^2, \ldots$. The p-adic Gauss mean for f/p is the same as that for

$$[b_1, b_2, \ldots, pc_1, pc_2, \ldots, p^2 d_1, \ldots]$$

since that for $[a_i/p]$ is 1.

The means for f/p^2, f/p^3, ... are the same as those for the forms

$$f' = [b_1, b_2, \ldots, pc_1, pc_2, \ldots]$$
$$f'' = [c_1, c_2, \ldots, pd_1, \ldots]$$
$$f''' = [d_1, \ldots]$$

$$\cdots$$

in a similar way. So for $p \geq 3$ we can recover the values of e_p and $\det_p$ for all the forms f', f'', f''', which is enough to determine the p-adic symbol.

To find the q-dimensions, we do this: we find, for instance,

$$\det_p(f)/\det_p(f') = p^{n_1 + n_p + \cdots}$$
$$\det_p(f')/\det_p(f'') = p^{n_p + n_{p^2} + \cdots},$$

from which we can recover n_1. In a similar way, we can recover the individual q-dimensions and p-excesses $e_p(f_1)$, $e_p(f_p)$, ... We conclude that:

Theorem. *For odd p, the p-adic symbol is an audible invariant.*

In particular, it *is* an invariant!

The case $p = 2$

The analogous discussion for $p = 2$ is very similar, but has some new features. We consider first the 2-adic Gauss means for the 1-dimensional forms $[2^k]$. Since $e^{\pi i} = -1$, the 2-adic Gauss mean of the form $[1]$ is the mean of $(-1)^{m^2}$ over all integers m, which is zero. The same thing happens for $[k]$ for any odd number k. For the forms $[2k]$ $(k = 1, 3, 5, 7)$ we get the mean of i^{km^2}, or equally of

$$
\begin{aligned}
(k = 1): \quad &1, i, i^4 = 1, i^9 = i \quad \text{i.e.,} \quad (1 + i)/2 = \zeta^1/\sqrt{2} \\
(k = 3): \quad &1, -i, 1, -i \quad \text{i.e.,} \quad (1 - i)/2 = \zeta^{3+4}/\sqrt{2} \\
(k = 5): \quad &1, i, 1, i \quad \text{i.e.,} \quad (1 + i)/2 = \zeta^{5+4}/\sqrt{2} \\
(k = 7): \quad &1, -i, 1, -i \quad \text{i.e.,} \quad (1 - i)/2 = \zeta^7/\sqrt{2}.
\end{aligned}
$$

Similar calculations for the forms

$$
[2^n k], \qquad 2^n \begin{pmatrix} 0 & 1 \\ 1 & 0 \end{pmatrix}, \qquad 2^n \begin{pmatrix} 2 & 1 \\ 1 & 2 \end{pmatrix}
$$

suffice to prove the

Theorem. *The 2-adic Gauss mean of an integral form*

$$
f = f_1 + 2f_2 + 4f_4 + \cdots
$$

is

$$
0, \text{ if } f_1 \text{ has type I}
$$

and otherwise

$$
\frac{\zeta^{\mathrm{oddity}(f)}}{\sqrt{\det_2(f)}}.
$$

Just as in the case $p \geq 3$, we can obtain the same information for the forms

$$
\begin{aligned}
f' &= f_2 + 2f_4 + 4f_8 + \cdots, \\
f'' &= f_4 + 2f_8 + \cdots, \\
f''' &= f_8 + \cdots,
\end{aligned}
$$

$$
\cdots
$$

but when any of the Gauss means vanish, they take some information with them. However, we should not have expected to determine the oddities of the individual f_q, because they are not invariants of f. In fact it turns out that

Theorem. *Except possibly for the q-dimensions n_1, n_2, n_4, ..., the 2-adic Gauss means of f, $f/2$, $f/4$, ... determine all the information in the 2-adic symbol, up to the variations that are allowed by the oddity-fusion and sign-walking rules.*

We shall see in a moment that the q-dimensions are not always audible, so that we cannot guarantee recovering them from the Gauss means. However, they are certainly invariants of f—indeed, 2-adic invariants—since they specify the structure of the 2-part of the dual quotient group L^*/L, which is the direct product of

n_1 copies of C_1, n_2 copies of C_2, n_4 copies of C_4, ...

So we can still assert

Theorem. *The 2-adic symbol (up to oddity-fusion and sign-walking) is an invariant of f under 2-adic integral equivalence.*

Hearing the genus: a hide-and-seek game

We now know that we can hear all of the genus except possibly the q-dimensions n_q for $p = 2$. Some information about these is audible, however, and we shall show that this will enable us to hear all these dimensions when n is 4.

We do this by playing a game against a team of n players, who distribute themselves among various houses (n_q of them going into house H_q) according to certain rules. Each house must either display one of the players in it (this is the type I case, and the requirement corresponds to the fact that n_q must be positive when f_q has type I) or display a card bearing the number

$$N_q = n_{2q} + 2n_{4q} + 3n_{8q} + 4n_{16q} + \cdots,$$

which is the sum of the distances from H_q of all players in houses to the right of H_q (this is the type II case, and the determinant of $f_q + 2f_{2q} + 4f_{4q} + \ldots$ is 2^{N_q}). Moreover, the number of players in any type II house must be even (it may be 0).

For example, the figure below displays the first four houses of a form with f_1 and f_8 of type I and f_2 and f_4 of type II.

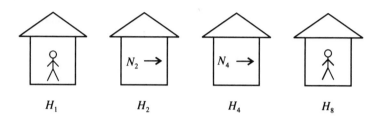

$$H_1 \qquad H_2 \qquad H_4 \qquad H_8$$

We shall show that if the opposing team has just four players, we can find where they are. If all the houses are type II, we have complete information, so we can suppose that at least one of them is type I, and we can therefore see one player, in a house H_q say. If we cannot see another player, all houses other than H_q must have an even number of players (since they are type II), so H_q must also have an even number of players, and we have located a second player.

Now suppose that there are two possibilities for the locations of the remaining two players that are compatible with all the information we can see. Then we can imagine moving two players from houses H_B and H_C to houses H_A and H_D, say, without changing any of this information. Without loss of generality, we can suppose that

$$H_A < H_B \leq H_C < H_D,$$

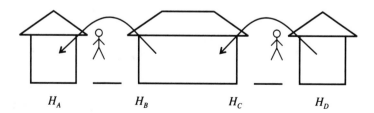

$$H_A \qquad H_B \qquad H_C \qquad H_D$$

and we must have $BC = AD$ since the determinant must be unaltered. Now each of H_A and H_D has an odd number of players in one of these configurations, so must be type I, and the same must be true of H_B and H_C unless they are the same house. So if we can't see a third player, then H_B and H_C must be the same type II house—but then the number on the card displayed for this house would be different for the two configurations. We can therefore locate a third player in all cases, and the location of the fourth player can now be found from the determinant.

We have proved a theorem that I believe is due to Kitaoka:

Theorem. *The genus of a quadratic form of dimension 4 or less is an audible invariant.*

Inaudibility of the genus in higher dimensions

The cubic and isocubic lattices of minimal norm 2 that were mentioned in the Second Lecture have distinct 2-adic symbols:

$$1_\infty^0 2_6^6 4_\infty^0 \qquad \text{for the cubic lattice}$$
$$1_\infty^2 2_6^2 4_\infty^2 \qquad \text{for the isocubic lattice}$$

The sublattices of these perpendicular to the vector $(1,1,1,1,1,1)$ also have distinct 2-adic symbols:

$$1_\infty^0 2_3^{-5} 4_\infty^0,$$
$$1_\infty^2 2_3^{-1} 4_\infty^2.$$

We deduce:

Theorem. *The genus is not always audible when the dimension is 5 or more.*

We believe that this result appears for the first time in this book.

It is interesting to see how the opposing team defeats us with six players. The cubic and isocubic lattices both give the picture below:

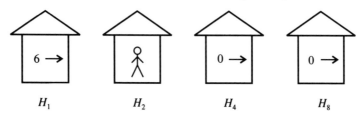

but for the cubic lattice all six players are in H_2, while for the isocubic one there are two players in each of H_1, H_2, H_4. In each case the sum of all the distances from H_1 of players to the right of H_1 is 6. For the 5-dimensional examples we remove one player from H_2.

More About the Invariants: The p-Adic Numbers

Invariance of the p-signatures

We have not yet proved that the p-signatures (or equivalently the p-excesses) are invariants of rational equivalence. We can do this by showing that for any type II integral form F that is rationally equivalent to f, the p-adic Gauss mean of F is $\zeta^{\sigma_2(f)}$ or $\zeta^{-e_p(f)}$ times a real positive number, according as p is or is not 2.

If F_1 and F_2 are two type II integral forms equivalent to f, then we can get from one to the other by a chain of steps, each of which replaces a lattice by a sublattice of prime index P in it, or vice versa.

Now for primes p other than P, such a replacement merely multiplies the terms in the Gauss mean by Pth roots of unity, and so does not affect the p-adic Gauss mean or the p-excess. But e_P can be computed from the other e_p by the global relation proved in the Postscript, and so it is fixed too!

The p-adic numbers

Gauss and others noticed very early on that some of the most interesting properties of quadratic forms depend only on the values of their coefficients modulo various powers of primes. K. Hensel found a very nice way to express these by introducing his rings of p-adic numbers.

117

We shall define a p-*adic integer* x to be the formal solution of a system of consistent congruences

$$x \equiv a_n \pmod{p^n}, \qquad n = 0, 1, 2, \ldots,$$

there being one congruence for each power of p. The consistency condition is that any two of these congruences should have a common solution in the ordinary integers. So for example, the congruences $x \equiv 1 \pmod 3$, $x \equiv 7 \pmod 9$, and $x \equiv -2 \pmod{27}$ are consistent, since $x = 25$ satisfies all of them.

These p-adic integers form a ring, since they can be added, subtracted, and multiplied. Indeed, if x and y are the two p-adic integers defined by the congruences

$$x \equiv a_n \pmod{p^n}, \qquad y \equiv b_n \pmod{p^n},$$

then $u = x + y$, $v = x - y$, $w = xy$ are the p-adic integers defined by

$$u \equiv a_n + b_n, \ v \equiv a_n - b_n, \ w \equiv a_n b_n \pmod{p^n}.$$

The number $1/3$ is a 2-adic integer; also a 5-adic integer. To see this, observe that the following numbers a_n

$$7, 67, 667, 6667, 66667, \ldots$$

multiply by 3 to give the numbers

$$21, 201, 2001, 20001, 200001, \ldots,$$

which are congruent to 1 modulo higher and higher powers of 2 and 5. So the p-adic number x defined by $x \equiv a_n \pmod{p^n}$, where p is 2 or 5, satisfies $3x = 1$.

Again, there is a 5-adic integer x such that $x^2 = -1$, since the numbers

$$2, 7, 57, 182, \ldots$$

(this time the general rule is less obvious) are congruent to each other, and their squares

$$4, 49, 3249, 33124, \ldots$$

to -1, modulo increasing powers of 5.

More generally, if r/s is any rational number whose denominator is not divisible by p, then there is a congruence $x \equiv a_n \pmod{p^n}$ that is equivalent to $sx \equiv r$, and the formal solution of all these congruences is a p-adic integer x satisfying $sx = r$, which it is natural to identify with the rational number r/s. And if k is any quadratic residue modulo p, we can show in a similar way that the equation $x^2 = k$ is solvable in the p-adic integers.

The *p-adic rationals* are defined in just the same way as the p-adic integers, but now in the defining congruences

$$x \equiv a_n \pmod{p^n},$$

the numbers a_n may be arbitrary rational numbers. It turns out that every p-adic rational is a p-adic integer divided by some power of p.

We shall use $\mathbf{Z}_p$ for the set of p-adic integers, and $\mathbf{Q}_p$ for the set of p-adic rationals. All this discussion has been for primes other than the rather special prime -1. There are alternative approaches which make it natural to define both $\mathbf{Z}_{-1}$ and $\mathbf{Q}_{-1}$ to be the ring $\mathbf{R}$ of real numbers.

The reader who is unfamiliar with these notions will find that they are not really needed in our book. Little is lost when a phrase such as

"*the equation ...... is solvable in the p-adic integers*"

is replaced by

"*the congruence ...... is solvable modulo arbitrarily high powers of p*".

The binary forms over $\mathbf{Q}_p$

Over any of the fields $\mathbf{Q}$ or $\mathbf{Q}_p$, a non-singular binary form of determinant d is completely determined by giving any particular number that it represents, for if it represents $k \neq 0$, it is equivalent to $[k, dk]$, and if it represents 0, it is equivalent to the form xy, or $[1, -1]$.

When a form represents k, it also represents all the numbers kx^2 ($x \neq 0$), which constitute the *squareclass* of k. Our names for the

p-adic squareclasses of nonzero numbers are

$$
\begin{array}{ll}
+u, -u & \text{for } p = -1, \\
u_1, u_3, u_5, u_7, 2u_1, 2u_3, 2u_5, 2u_7 & \text{for } p = 2, \\
u_+, u_-, pu_+, pu_- & \text{for } p \geq 3,
\end{array}
$$

where u is the class of positive numbers, u_k is the 2-adic squareclass containing numbers of the form $8n + k$ ($k = 1, 3, 5, 7$), and u_+, u_- are the p-adic squareclasses containing the quadratic residues (i.e., squares) and non-residues, respectively.

Since there are only finitely many squareclasses over $\mathbf{Q}_p$, there are only finitely many forms of any given dimension. The binary ones are classified by the squareclasses of their determinants and the numbers they represent (at nonzero vectors), in the following table (in which $p_\pm$ denotes a positive prime congruent to ± 1 modulo 4):

p	d	numbers represented by the two forms	
-1	$+u$	$\{+u\}$	$\{-u\}$
	$-u$	$\{0, +u, -u\}$	
2	u_1	$\{u_1, u_5, 2u_1, 2u_5\}$	$\{u_3, u_7, 2u_3, 2u_7\}$
	u_3	$\{u_1, u_3, u_5, u_7\}$	$\{2u_1, 2u_3, 2u_5, 2u_7\}$
	u_5	$\{u_1, u_5, 2u_3, 2u_7\}$	$\{u_3, u_7, 2u_1, 2u_5\}$
	u_7	$\{0, u_1, u_3, u_5, u_7, 2u_1, 2u_3, 2u_5, 2u_7\}$	
	$2u_1$	$\{u_1, u_3, 2u_1, 2u_3\}$	$\{u_5, u_7, 2u_5, 2u_7\}$
	$2u_3$	$\{u_1, u_7, 2u_3, 2u_5\}$	$\{u_3, u_5, 2u_1, 2u_7\}$
	$2u_5$	$\{u_1, u_3, 2u_5, 2u_7\}$	$\{u_5, u_7, 2u_1, 2u_3\}$
	$2u_7$	$\{u_1, u_7, 2u_1, 2u_7\}$	$\{u_3, u_5, 2u_3, 2u_5\}$
p_+	u_+	$\{0, u_+, u_-, pu_+, pu_-\}$	
	u_-	$\{u_+, u_-\}$	$\{pu_+, pu_-\}$
	pu_+	$\{u_+, pu_+\}$	$\{u_-, pu_-\}$
	pu_-	$\{u_+, pu_-\}$	$\{u_-, pu_+\}$
p_-	u_+	$\{u_+, u_-\}$	$\{pu_+, pu_-\}$
	u_-	$\{0, u_+, u_-, pu_+, pu_-\}$	
	pu_+	$\{u_+, pu_+\}$	$\{u_-, pu_-\}$
	pu_-	$\{u_+, pu_-\}$	$\{u_-, pu_+\}$

Rational forms with prescribed invariants

The reader will have noticed from the above table that whenever d is in the square-class of -1, there is only one binary form of determinant d, and that this form represents all numbers. These are the *binary isotropic forms*. More generally, any nonsingular form that represents 0 non-trivially is called *isotropic*, and we can easily prove that the isotropic forms are precisely those of the shape $[1, -1, *, *, \ldots]$. For, on the one hand, the form $[1, -1]$ is isotropic; and on the other hand if $[a, b, \ldots]$ is isotropic, there is an equation $ax^2 + by^2 + \ldots = 0$, in which there must be at least two non-zero terms. But then the vector $\mathbf{v}$ for which $f(\mathbf{v}) = 0$ can be written as the sum of two orthogonal vectors of nonzero norms k and $-k$, and so with respect to a suitable base, f looks like $[k, -k, *, *, \ldots] = [1, -1, *, *, \ldots]$.

So over Q_p there are two binary forms of determinant d, unless $-d$ is a p-adic square, when there is only the isotropic form. Now given a rational number d, choose one of the two forms $f^{(p)}$ of determinant d for each p (except that of course if $-d$ is a p-adic square there is no choice). Under what conditions does there exist a rational quadratic form of determinant d that is p-adically equivalent to $f^{(p)}$ for each p?

The answer is that such an f exists just if the global relation holds—the sum of the p-excesses of all the $f^{(p)}$ must be 0 mod 8.

Legendre's proof of the Three Squares Theorem used at a corresponding point the fact that suitable arithmetic progressions contain prime numbers, which was only proved by Dirichlet many years later. (Serre, in his *Cours d'Arithmetique* [Ser], also uses the Dirichlet theorem for this result—but unlike most other authors he later proves that theorem!) [3]We shall follow the tradition of quoting the Dirichlet

---

[3] Gauss found a proof independent of the Dirichlet theorem using his composition law and the genera of binary quadratic forms over **Z**. We briefly sketch the Gauss proof, for those who understand the terms. We want to show that the genus has "the right size". But two forms are in the same genus just if their quotient is a square in the composition group. So we can alternatively show that the kernel of the square map has the right size. The forms in the kernel are well-known—they are the so-called *ambiguous* forms; they correspond roughly to the factorizations of d, and there are the right number of them.

theorem, and will also use the global relation, which we shall only prove in the Postscript.

We briefly sketch the production of the form f from the forms $f^{(p)}$. In the binary case, we take the desired form to be

$$[p_1 p_2 p_3 \cdots p_m P, \; p_1 p_2 \cdots p_m P d]$$

where the primes $p_1, p_2, \ldots p_m$ are selected from -1, 2, the primes dividing d, and any others for which the desired p-adic form $f^{(p)}$ is not equivalent to $[1, d]$. The number P is a large prime to be determined. Now it is easy to see that we can control $f^{(p)}$ by deciding whether or not to include p in the product $p_1 p_2 \cdots p_m$, and at the same time adjusting the quadratic residuacity of P modulo p.

This means that we can get the correct form $f^{(p)}$ for all primes p except possibly P. But the global relation then ensures that $f^{(P)}$ will be correct, too!

To find a higher dimensional form

$$f = [a_1, a_2, a_3, \ldots]$$

with prescribed invariants, we first choose values for a_3 onwards in such a way that the value desired for the product $a_1 a_2$ is not the negative of a p-adic square for any relevant p, and then use the above method to find the binary subform $[a_1, a_2]$.

Integral forms with prescribed invariants

A slightly more cumbrous argument can be used to produce an integral quadratic form with prescribed p-adic symbols, provided that these satisfy the global relation. Once again, the easiest proof uses Dirichlet's theorem, but this can be avoided at the cost of extra work. The form can be made tridiagonal: all nonzero matrix entries are on the principal diagonal or one of the two adjacent ones. The leading matrix entry (in the top left corner) can be chosen to be any number that is primitively represented over the p-adic integers for every prime p. Then subsequent entries are selected in sequence modulo powers of various primes, using the Chinese Remainder Theorem. At most one new prime need be introduced at each stage, using Dirichlet's theorem.

We cannot control the behavior with respect to the last such prime, but the global relation deals with it automatically.

This has an important consequence for the representation theory of integral forms. If an integer n is represented by f p-adically for each p (including -1), then although it may not be represented by f itself, it *is* represented by some form in the same genus as f. The reason is that if n is primitively represented p-adically for each p, the above argument constructs a form g in the same genus as f, and whose leading entry is n, so that g primitively represents n.

Of course f represents n imprimitively just if it represents some integer n/k^2 ($k > 1$) primitively. If this happens p-adically for each p, then some g in the same genus as f represents n/k^2 primitively and so represents n imprimitively.

This is what really underlies Legendre's Three Squares Theorem. There is only one form in the genus of $f(x, y, z) = x^2 + y^2 + z^2$, so if a number satisfies the p-adic conditions for representability by f, it must actually be represented by f itself. The only non-trivial conditions are for $p = -1$ and 2. However, we shall give a complete proof of Legendre's Theorem, not involving this principle, in the Postscript.

Equivalence with inessential denominator

The orthogonal group for a quadratic form f is generated by reflections in various vectors $\mathbf{v}$, with $f(\mathbf{v}) \neq 0$. Let $\{-1, 2, \ldots, p\}$ be a finite set of primes which includes -1, 2 and all primes dividing $\det(f)$; we shall call these the *essential primes*. The Approximation Theorem says that if we are given p-adically integral automorphisms

$$\theta^{(-1)}, \theta^{(2)}, \ldots, \theta^{(p)}$$

of f of the same determinant $+1$ or -1, for all essential primes, then there is a rational automorphism θ which localizes to these $\theta^{(p)}$.

To prove this, note that we can write $\theta^{(p)}$ as the product of reflections in vectors

$$\mathbf{r}_1^{(p)}, \mathbf{r}_2^{(p)}, \ldots, \mathbf{r}_k^{(p)}$$

where k is whichever of n and $n - 1$ has $(-1)^k$ equal to the desired determinant. Then take θ to be the product of reflections in rational vectors

$$\mathbf{r}_1, \mathbf{r}_2, \ldots, \mathbf{r}_k$$

that are congruent to the above vectors modulo sufficiently high powers of the appropriate p.

What we have just proved is the principle of "equivalence with inessential denominator"—if two forms are in the same genus, then they are equivalent under a rational transformation, the denominators of whose matrix entries involve no essential primes.

The spinor genus

There are actually two interesting homomorphisms defined on the rational orthogonal group of a quadratic form f, the *determinant*, with values ± 1, and the *spinor norm*, whose values are rational square-classes. They are well enough defined by saying that the reflection in a vector $\mathbf{r}$ has determinant -1 and spinor norm equal to the square-class of $f(\mathbf{r})$.

The spinor genus is a refinement of the genus using these notions. If f and g are in the same genus, they are equivalent by many different rational transformations with inessential denominator.

Let $S_r(f)$ be the class of all g for which such a rational transformation can be chosen to have determinant 1 and spinor norm r. Then $S_1(f)$ is the *spinor genus* of f, and the *spinor kernel* is defined to be the set of all numbers r for which $S_r(f) = S_1(f)$.

Our definition of the spinor genus seems to be the simplest one. It is midway between the traditional one (which is for a slight modification of Eichler's original concept) and Watson's later definition of "spinor equivalence".

We can readily compute with it using Watson's

Theorem. *Let f and g be quadratic forms of the same determinant d corresponding to lattices L and M whose intersection has index* *r in each of them. Then if r is an odd number prime to d, we may conclude that g is in $S_r(f)$. In particular, f and g will have the same genus, and also the same spinor genus if r is in the spinor kernel.*

Its importance arises from Eichler's

Theorem. *Two indefinite forms of rank at least 3 are integrally equivalent just if they lie in the same spinor genus.*

The spinor genus as we have defined it is an effectively computable invariant. In practice one works with the spinor kernel, which can be computed by purely "local" calculations.

Very often the spinor genus coincides with the genus, so that our p-adic symbols characterize the form up to equivalence. Indeed, it can be shown that this happens unless for some p the form can be diagonalized and the powers of p that divide each entry are distinct. Also, for indefinite forms of dimension n and determinant d, these two concepts coincide unless

$$4^{[\frac{n}{2}]}d \text{ is divisible by } k^{\binom{n}{2}}$$

for some nonsquare natural number $k \equiv 0$ or 1 modulo 4. This is Theorem 21 of Chapter 15 of [CS], where there is also given a practicable algorithm for computing with the spinor genus.

A Taste of Number Theory

Three famous theorems

In this Postscript we shall prove three famous theorems. These are the notorious quadratic reciprocity law, the fact that the signature of an even unimodular quadratic form is a multiple of 8, and Legendre's celebrated three squares theorem. We shall derive some consequences of Legendre's theorem, including the universality of certain forms in four variables, and finish by explaining why no rational positive definite ternary form is universal.

Zolotarev's definition of the Jacobi symbol

For an odd number n whose prime factorization is $pqr \ldots$, Jacobi defined his symbol to be

$$\left(\frac{a}{n}\right) = \left(\frac{a}{p}\right)\left(\frac{a}{q}\right)\left(\frac{a}{r}\right)\ldots,$$

a product of Legendre symbols. Although it was clear from its many properties that the Jacobi symbol was a very natural object, it was some time before Zolotarev found a more meaningful definition: $\left(\frac{a}{n}\right)$ is the sign of the permutation obtained by multiplying by a modulo n. In this chapter we shall adopt Zolotarev's definition.

We recall that every permutation π of a finite set has a *sign*, which is -1 just if an odd number of the cycles in π have even length. Now we shall write "$\times a \bmod n$" for the permutation of $\{0, \ldots, n-1\}$ corresponding to multiplication by a. Thus

$$\times 3 \bmod 11 \; = \; (0)(1,3,9,5,4)(2,6,7,10,8).$$

Since this has no even length cycle, $\left(\frac{3}{11}\right) = +1$.

This definition leads to an extremely simple proof of the quadratic reciprocity theorem. It is remarkable that this proof does not use either the notion of prime number, or even that of square number. We shall however use the fact that the sign of a permutation is multiplicative.

We shall say that a number is *positive modulo m* just if it is congruent modulo m to something strictly between 0 and $\frac{1}{2}m$ and *negative modulo m* if instead it is congruent to something strictly in the range $(-\frac{1}{2}m, 0)$. The number $\frac{1}{2}m$ is *ambiguous modulo m*.

Five lemmas

We first evaluate $\left(\frac{-1}{n}\right)$.

Lemma 1. $\left(\frac{-1}{n}\right)$ *is the sign of n modulo* 4.

In other words, $\left(\frac{-1}{n}\right)$ is 1 if n is $4k + 1$ and -1 if n is $4k - 1$.

Proof.   This is immediate from the definition. For example, $\left(\frac{-1}{11}\right)$ and $\left(\frac{-1}{13}\right)$ are the signs of the permutations

$$(0)(1, -1) \ldots (-5, -5) \text{ and } (0)(1, -1) \ldots (6, -6),$$

namely -1 and 1, since these have, respectively, 5 and 6 transpositions. $\square$

Lemma 2.   *We have $\left(\frac{a}{n}\right) = (-1)^s$, where s is the "sign change number" for $\times a \bmod n$, namely the number of positive numbers k mod n for which ak is negative mod n.*

Proof. We shall consider the case $\left(\frac{3}{11}\right)$. For this we shall display the numbers permuted by the permutation $\times 3$ mod 11, with their proper signs:

$$(0)(+1, +3, -2, +5, +4)(-1, -3, +2, -5, -4).$$

We analyze this in the figure below which manifests the sign change number $s = 2$ as the number of crossing points. The figure shows that this permutation factors as the product of the two permutations

$$(0)(1, 3, 2, 5, 4)(-1, -3, -2, -5, -4) \text{ and } (2, -2)(5, -5).$$

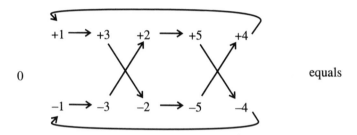

equals

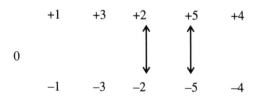

times

The first of these is the permutation we might call "absolute multiplication by 3," obtained by multiplying by 3 except that one preserves the sign, while the second consists of the necessary sign corrections.

In the general case, absolute multiplication by a mod n is an even permutation, since it is the product of two permutations of exactly the same shape, while the sign correction permutation consists of s transpositions, and thus has sign $(-1)^s$.   □

For any fixed a, the symbol $\left(\frac{a}{n}\right)$ can be evaluated for all n using this lemma; for example, $\left(\frac{2}{n}\right) = 1$ or -1 according as $n \equiv \pm 1$ or ± 3 modulo 8.

Lemma 3. *If $a > 0$, then s is the number of integers strictly between 0 and $n/2$ that lie in intervals of the form*

$$\left[\left(l - \frac{1}{2}\right) \frac{n}{a}, \; l\frac{n}{a} \right].$$

Proof.

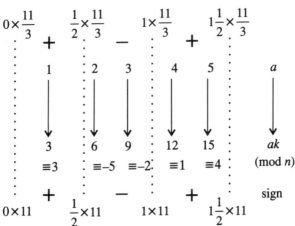

The picture makes this immediate. The dotted lines show how the sign changes from $+$ to $-$ when ak passes a number of the form $(l - \frac{1}{2})n$ and back from $-$ to $+$ as ak passes a number of the form ln.   □

Lemma 4. (Periodicity of the Jacobi symbol in n)
If $m \equiv \pm n \pmod{4a}$, then $\left(\frac{a}{m}\right) = \left(\frac{a}{n}\right)$.

Proof. If we add or subtract a multiple of $4a$ to n, all the endpoints of the intervals in Lemma 3 change by even integers, so s changes by an even integer. Also the symbol $\left(\frac{a}{-n}\right)$ equals $\left(\frac{a}{n}\right)$, since multiplication by a modulo $-n$ is the same as multiplication by a modulo n. $\square$

Lemma 5. *If m and n are two coprime odd integers whose sum is a positive multiple of 4, then $\left(\frac{m}{n}\right) = \left(\frac{n}{m}\right)$.*

Proof. Write $m + n = 4a$. Then

$$\left(\frac{m}{n}\right) = \left(\frac{4a}{n}\right) = \left(\frac{a}{n}\right) = \left(\frac{a}{m}\right) = \left(\frac{4a}{m}\right) = \left(\frac{n}{m}\right).$$

This is because to multiply by m modulo n is the same as to multiply by $4a$ modulo n, which we can do by multiplying by 2 twice (which does not affect the sign) and then multiplying by a. $\square$

Reciprocity for the Jacobi symbol

The usual formulation of the reciprocity law for the Jacobi symbol is

Theorem. *If m and n are two positive coprime odd numbers, then $\left(\frac{n}{m}\right) = \left(\frac{m}{n}\right)$ **unless** m and n are both negative modulo 4, in which case $\left(\frac{m}{n}\right)$ and $\left(\frac{n}{m}\right)$ are distinct.*

Proof. It is easy to see that Lemma 5 contains this assertion. We discuss three cases using the fact that $\left(\frac{m}{-n}\right) = \left(\frac{m}{n}\right)$ (since multiplication modulo $-n$ is the same as multiplication modulo n):

1) *Opposite signs modulo 4.* In this case, Lemma 5 gives the theorem directly. For example, $\left(\frac{11}{13}\right) = \left(\frac{13}{11}\right)$ because $11 + 13$ is divisible by 4.

2) *Both positive modulo* 4. We consider 5 and 13. Here, Lemma 5 shows that

$$\left(\frac{13}{5}\right) = \left(\frac{13}{-5}\right) = \left(\frac{-5}{13}\right) = +\left(\frac{5}{13}\right).$$

because $13 - 5$ is a positive multiple of 4, and $\left(\frac{-1}{13}\right) = +1$ by Lemma 1.

3) *Both negative modulo* 4. We consider 7 and 11, and apply Lemma 5 to find

$$\left(\frac{11}{7}\right) = \left(\frac{11}{-7}\right) = \left(\frac{-7}{11}\right) = -\left(\frac{7}{11}\right)$$

since $11 - 7$ is a positive multiple of 4, and $\left(\frac{-1}{11}\right) = -1$, by Lemma 1.

This proof has removed much of the mystery from the reciprocity law by using Zolotarev's Jacobi symbol rather than the Legendre symbol. Otherwise, it is patterned on a proof of Scholz from 1939 [Scho].

Legendre symbols and linear Jacobi symbols

To show that this reciprocity law implies the usual quadratic reciprocity law, we must show that the Jacobi symbol $\left(\frac{a}{p}\right)$ as defined by Zolotarev coincides with the traditional Legendre symbol when p is an odd prime.

To do this, we quote Euler's theorem that the multiplicative group of integers modulo p is a cyclic group of order $p - 1$. But if g is a generator for this group, then the permutation "$\times g \bmod p$" is a $(p-1)$-cycle, and so $\left(\frac{g^k}{p}\right) = (-1)^k$, establishing the desired result.

One traditional way to express the reciprocity law is

$$\left(\frac{m}{n}\right)\left(\frac{n}{m}\right) = (-1)^{\frac{m-1}{2}\frac{n-1}{2}}$$

provided m and n are coprime.

To simplify our work with p-signatures and p-excesses, it will be convenient to introduce a linear version of the Jacobi symbol, whose value is an integer modulo 8. We define

$$\begin{bmatrix} m \\ \overline{} \\ n \end{bmatrix} \qquad \text{to be} \qquad 0 \quad \text{or} \quad 4$$

according as

$$\left(\frac{m}{n} \right) \qquad \text{is} \qquad -1 \quad \text{or} \quad 1.$$

The reciprocity law now becomes

$$\begin{bmatrix} m \\ \overline{n} \end{bmatrix} + \begin{bmatrix} n \\ \overline{m} \end{bmatrix} \equiv (m-1)(n-1) \pmod 8.$$

The global relation

In the fourth lecture, we defined invariants σ_p, called the p-signatures, of a rational quadratic form, whose values are integers or integers modulo 8. We also defined the *p-excesses* e_p by

$$e_p(f) = \sigma_p(f) - \dim(f) \qquad \text{if } p \neq 2,$$

and

$$e_2(f) = \dim(f) - \sigma_2(f).$$

Now since each e_p can be computed just from the p-adic version of the form, one would naturally expect the e_p's to be independent. However, there is a remarkable global relationship:

The sum of all the p-excesses is a multiple of 8.

The same relation has usually been expressed by saying that the product of certain invariants is 1, and so it is usually known as "the product formula". The p-signature for the 1-dimensional form $[a] = [p^\alpha A]$ was defined in the fourth lecture to be

$$p^\alpha \text{ or } p^\alpha + 4,$$

the 4 being present when a is a p-adic antisquare, that is to say, when α is odd and A is a quadratic nonresidue modulo p. In terms of our linear Jacobi symbol, this becomes

$$p^{\alpha} + \left[\frac{A}{p^{\alpha}}\right].$$

This can be generalized! If π is any set of odd primes, then we shall define the π-*signature* of $[a]$ to be

$$\pi(a) + \left[\frac{\pi'(a)}{\pi(a)}\right],$$

where $\pi(a)$ and $\pi'(a)$ are the portions of a composed of primes in π and in the complementary set π', respectively. The π-*excess* e_{π} of $[a]$ is defined to be 1 less, namely:

$$\pi(a) - 1 + \left[\frac{\pi'(a)}{\pi(a)}\right].$$

The π-signature and π-excess of a general form $f = [a, b, c, \ldots]$ are then defined to be the sums

$$\sigma_{\pi}(f) = \sigma_{\pi}(a) + \sigma_{\pi}(b) + \cdots$$

and

$$e_{\pi}(f) = e_{\pi}(a) + e_{\pi}(b) + \ldots,$$

so that $e_{\pi}(f) = \sigma_{\pi}(f) - \dim(f)$.

Theorem. *Let π_1 and π_2 be two disjoint sets of odd primes, and π their union. Then $e_{\pi} \equiv e_{\pi_1} + e_{\pi_2} \pmod{8}$.*

Proof. Let $a = P_1 P_2 A$, where P_1 and P_2 are the π_1 and π_2 parts of a. Then

$$e_{\pi_1}(a) = (P_1 - 1) + \left[\frac{P_2}{P_1}\right] + \left[\frac{A}{P_1}\right]$$
$$e_{\pi_2}(a) = (P_2 - 1) + \left[\frac{P_1}{P_2}\right] + \left[\frac{A}{P_2}\right]$$
$$e_{\pi}(a) = (P_1 P_2 - 1) + \left[\frac{A}{P_1}\right] + \left[\frac{A}{P_2}\right].$$

So the theorem follows immediately from the reciprocity law

$$(P_1 - 1)(P_2 - 1) \equiv \left[\frac{P_1}{P_2}\right] + \left[\frac{P_2}{P_1}\right] \quad (\text{mod } 8).$$

What about sets of primes containing 2? We can make the additivity result continue to hold for these sets merely by defining e_π to be $-e_{\pi'}$ for any such sets. This is consistent with the above definition of the 2-excess.

We have now proved our global relation:

$$e_{-1}(a) + e_2(a) + e_3(a) + e_5(a) + \ldots \equiv 0 \quad (\text{mod } 8),$$

for 1-dimensional forms $[a]$, and by additivity we can deduce it for all forms f. The sum is really a finite one because all but finitely many terms are 0.

The strong Hasse-Minkowski principle

The global relation together with the existence of rational forms with prescibed invariants can be used to prove an important principle sometimes called the *strong Hasse-Minkowski principle*.

Theorem. *A rational form f represents a rational number $r \neq 0$ over the rational field* $\mathbf{Q}$ *if and only if it represents r over every* $\mathbf{Q}_p$ *($p = -1, 2, 3 \ldots$).*

Proof. The number r is represented by f over some field just if f is equivalent to a form $[r, *, *, \ldots]$ over that field, or in other words, just if f is equivalent to $[r] \oplus g$ over that field, for some g. So the hypotheses give us forms $g^{(p)}$ for which f is p-adically equivalent to $[r] \oplus g^{(p)}$ and the conclusion requires a form g for which f is rationally equivalent to $[r] \oplus g$.

The p-adic invariants of the desired form g must therefore be the same as those of the given forms $g^{(p)}$. Does there exist such a g? Yes! The only condition is the global relation, and this holds for g because it holds for f and for $[r]$!   □

A theorem on even unimodular forms

For $p \geq 3$, an integral quadratic form whose determinant is prime to p has a p-adic integral diagonalization

$$f \cong [a_1, a_2, \ldots, a_n]$$

in which all of the a_i are prime to p. So its p-signature is n and its p-excess is 0. For a unimodular form, the global relation therefore simplifies to the assertion that $e_{-1}(f) + e_2(f) \equiv 0 \pmod 8$, or equivalently that the signature of f is congruent to its oddity modulo 8. We shall now prove that an *even* unimodular quadratic form has zero oddity, and so zero signature mod 8.

Such a form necessarily represents a number $2a$ for which a is odd, and we can use this inductively to transform it into a direct sum of forms of the shape:

$$\begin{pmatrix} 2a & b \\ b & 2c \end{pmatrix} \quad a, b \text{ odd}$$

over the 2-adic integers. The displayed form is equivalent over the 2-adic rationals to

$$[2a, 2a'] \qquad a' = ad,$$

where d is the determinant $4ac - b^2$ of the form, which is congruent to -1 mod 4. Thus one of a, a' is 1 or 5 (mod 8), and the other is 3 or 7 (mod 8). It follows that the oddity of this form is

$$(1 \text{ or } 5 + 4) \quad + \quad (3 + 4 \text{ or } 7),$$

which in each case is a multiple of 8.

The history of the even unimodular lattices

The remarkable fact that the dimension of an even unimodular lattice is a multiple of 8 is the positive definite case of the above result. We have seen that this is a consequence of the product formula. On the other hand, it can be shown that just as in the rational case, the global relation (which is an almost immediate consequence of quadratic reciprocity) is the only relation between the p-adic structures for each p, and so when it is satisfied, there exists a quadratic form over the rational integers with prescribed p-adic invariants. Indeed, this result is quantized by the mass formula, which in a suitable sense counts the number of integral quadratic forms with prescribed invariants.

In 1867, H.J.S. Smith [Smi] applied the mass formula to show that there exists an 8-dimensional even unimodular lattice, and this lattice E_8 was explicitly constructed by Korkine and Zolotareff in 1873. The mass formula can also be used to show that this lattice is unique. Witt in 1941 found the two such lattices in 16 dimensions that we mentioned in the second lecture. In 1973, Niemeier showed that there were just 24 even unimodular lattices in 24 dimensions, including the notorious Leech lattice.

The Three Squares Theorem

In the next section, we shall prove Legendre's celebrated *Three Squares Theorem* of 1798: a positive integer is a sum of three integral square numbers if and only if it is not the product of a power of 4 and a number congruent to -1 modulo 8. In this section, we show that such an n is the sum of three rational squares.

Now a form represents a number over the rationals just if it does so over each $\mathbf{Q}_p$. So n is the sum of three rational squares if and only if it can be represented as the sum of three real squares and as the sum of three p-adic rational squares for each positive prime $p = 2, \ldots$. This is very easy to decide since we only need discuss one number in each p-adic square-class.

We discuss the cases:

$p = -1$: -1 is clearly not representable, but $+1 = 1^2 + 0^2 + 0^2$ is. So the condition for a non-zero n to be the sum of three real squares is just that it be positive.

$p = 2$: The equations

$$1 = 1^2 + 0^2 + 0^2, \quad 3 = 1^2 + 1^2 + 1^2, \quad 5 = 2^2 + 1^2 + 0^2,$$
$$2 = 1^2 + 1^2 + 0^2, \quad 6 = 2^2 + 1^2 + 1^2, \quad 10 = 3^2 + 1^2 + 0^2,$$
$$14 = 3^2 + 2^2 + 1^2$$

handle all square-classes except that of -1 (or 7). We show on the other hand that -1 is not the sum of three squares of 2-adic rational numbers. If it were, then we could suppose that

$$-d^2 = a^2 + b^2 + c^2,$$

where a^2, b^2, and c^2 are all 2-adic integers, where at least one of them, a say, is odd (since otherwise we could cancel a factor of 2). Now, modulo 8, this congruence reads

$$-(0 \text{ or } 1 \text{ or } 4) \equiv 1 + (0 \text{ or } 1 \text{ or } 4) + (0 \text{ or } 1 \text{ or } 4),$$

and this is impossible.

$p \geq 3$:   As in the Fourth Lecture, we let $u = r + 1$ be the smallest positive number that is not a quadratic residue mod p, and suppose that $r \equiv x^2 \pmod{p}$. Then the square-classes of 1 and u are represented by sums of *two* squares, namely:

$$r \equiv x^2 + 0^2, \quad u \equiv x^2 + 1^2.$$

Any multiple mp of p is then a sum of three squares, since $mp - 1$ is in the same square-class as one of r or u and thus is the sum of two squares.

This discussion explains the role of the two conditions in Legendre's Theorem. The -1-adic condition is that n be positive; the 2-adic condition that it not be $4^a(8k - 1)$; the other p-adic conditions are always satisfied.

Representation by three integral squares

We shall show that any integer n that is the sum of three rational squares is actually also the sum of three integral squares, by using an ingenious reduction method published by Aubry in 1912 [Aub].

The equation

$$n = x_1^2 + x_2^2 + x_3^2$$

tells us that $n = (\mathbf{x}, \mathbf{x})$, where $\mathbf{x} = (x_1, x_2, x_3)$. Now if the x_i are not all integers, let m_i be the nearest integer to x_i, and let $x_i = m_i + r_i$. Then we have $\mathbf{x} = \mathbf{m} + \mathbf{r}$, where $0 < \mathbf{r} \cdot \mathbf{r} < 1$, and the vectors $\mathbf{m}$ and $d\mathbf{r}$ have integer coordinates. This shows that

$$n = (\mathbf{x}, \mathbf{x}) = (\mathbf{m}, \mathbf{m}) + 2(\mathbf{m}, \mathbf{r}) + (\mathbf{r}, \mathbf{r}),$$

which implies first that $2(\mathbf{m}, \mathbf{r}) + (\mathbf{r}, \mathbf{r})$ is an integer, N say, and then that $(\mathbf{r}, \mathbf{r})$ is a proper fraction with denominator d, say d'/d $(0 \leq d' < d)$. This in turn shows that the vector $\mathbf{r}/(\mathbf{r}, \mathbf{r}) = d\mathbf{r}/d'$ has denominator dividing d'.

Now reflect $\mathbf{x}$ in the vector $\mathbf{r}$! This produces a new vector $\mathbf{x}'$ with $(\mathbf{x}', \mathbf{x}') = (\mathbf{x}, \mathbf{x})$ and so a new representation of n as a sum of three rational squares. However, we find:

$$\begin{aligned}
\mathbf{x}' &= \mathbf{x} - 2(\mathbf{x}, \mathbf{r})/(\mathbf{r}, \mathbf{r})\mathbf{r} \\
&= \mathbf{m} + \mathbf{r} - 2(\mathbf{m}, \mathbf{r})/(\mathbf{r}, \mathbf{r})\mathbf{r} - 2\mathbf{r} \\
&= \mathbf{m} - N\mathbf{r}/(\mathbf{r}, \mathbf{r}),
\end{aligned}$$

which is a vector whose denominator divides d'.

This shows that from any representation of n as the sum of three squares of rational numbers with common denominator $d > 1$, we can derive another such representation with common denominator $d' < d$. Continuing in this way, we eventually find a representation of n with $d = 1$, that is, as a sum of three *integral* squares.

Consequences of Legendre's Theorem

Perhaps the most famous entry in the mathematical diary that Gauss kept as a young man is that for July 10, 1796, which reads

$$\text{EΥPHEKA!} \quad num \ = \ \Delta + \Delta + \Delta$$

Presumably Gauss had proved one of Fermat's assertions: every positive integer is the sum of three triangular numbers.

This follows easily from Legendre's theorem, which tells us that $8n + 3$ is the sum of three squares. But since these must all be odd, we have an equation

$$8n + 3 = (2a + 1)^2 + (2b + 1)^2 + (2c + 1)^2,$$

which entails

$$n = a(a + 1)/2 + b(b + 1)/2 + c(c + 1)/2.$$

The theorem also enables us to decide which numbers are sums of four *positive* squares. We first observe that numbers of the forms $8k + 3$ and $8k + 6$, since they are sums of three squares but not of two, are necessarily sums of three positive squares. Multiplying by 4, the same is true for numbers $32k + 12$ and $32k + 24$.

Theorem.   *The positive integers that are not the sums of four positive squares are precisely*

$$1, 3, 5, 9, 11, 17, 29, 41, \ 2 \times 4^m, \ 6 \times 4^m, \ 14 \times 4^m.$$

Proof.   We first show that any number greater than 49 that is not a multiple of 8 is the sum of four positive squares, by subtracting a square so as to yield a number of one of the forms described above. Thus from:

$8k + 2$ subtract 2^2 to get a number $8k + 6$
$8k + 3$ subtract 4^2 to get a number $8k + 3$
$8k + 4$ subtract 1^2 to get a number $8k + 3$
$8k + 6$ subtract 4^2 to get a number $8k + 6$
$8k + 7$ subtract 2^2 to get a number $8k + 3$
$8k + 1$ subtract 1^2 or 3^2 or 5^2 or 7^2 to get $32k + 24$
$8k + 5$ subtract 1^2 or 3^2 or 5^2 or 7^2 to get $32k + 12$.

The theorem is completed by checking the numbers up to 49, and verifying that a number $8k$ is the sum of four positive squares only if $2k$ is. [If 1 or 2 or 3 of the squares are odd, the number has form $4k + 1$ or $4k + 2$ or $4k + 3$; while if all four are odd, it has form $8k + 4$. So they must all be even.]   □

Similar methods show that various positive-definite forms in four variables are universal; that is to say, that they represent all positive integers. This holds for instance of the forms

$$x^2 + y^2 + z^2 + mt^2 \qquad \text{for } m = 1, 2, 3, 4, 5, 6, 7.$$

The Three Squares Theorem shows that if any such form misses any positive integer, then the smallest integer it misses must have the form $8k + 7$. But then by subtracting mt^2 for $t = 1, 1, 2, 1, 1, 1, 2$ in the seven cases, we obtain numbers of the respective forms $8k + 6$, $8k + 5$, $8k + 3$, $8k + 3$, $8k + 2$, $8k + 1$, $8k + 3$, which *are* represented by $x^2 + y^2 + z^2$.

The Fifteen Theorem

William Schneeberger and I have recently used these ideas to prove a remarkable theorem. If a positive definite quadratic form (in any number of variables) with integral matrix represents each of the numbers

$$1, \ 2, \ 3, \ 5, \ 6, \ 7, \ 10, \ 14, \ 15,$$

then it represents every positive integer. To see that this contains Lagrange's Four Squares Theorem, we just have to check that each of the above integers is the sum of at most four squares:

$$1 = 1^2, \quad 2 = 1^2 + 1^2, \quad 3 = 1^2 + 1^2 + 1^2, \quad 5 = 2^2 + 1^2,$$
$$6 = 2^2 + 1^2 + 1^2, \quad 7 = 2^2 + 1^2 + 1^2 + 1^2, \quad 10 = 3^2 + 1^2,$$
$$14 = 3^2 + 2^2 + 1^2, \quad 15 = 3^2 + 2^2 + 1^2 + 1^2,$$

and we are done!

The interested reader can use the theorem to check any other true assertion of this type; for instance, that every positive integer can be written as $a^2 + 2b^2 + 5c^2 + 5d^2 + 15e^2$.

The Fifteen Theorem is proved by arguments of the above type, but replacing the three squares form by various other forms g, found as follows. If f represents the above 9 numbers, then its lattice must contain vectors of norms $1, 2, 3, 5, \ldots$. There are only finitely many possibilities for the shape of the sublattice spanned by these vectors, and in almost all cases, we can find a 3-dimensional sublattice whose corresponding form g is unique in its genus. We then know just which numbers are represented by g, and we can use this to show that every number is represented by f.

No definite ternary form is universal

However, a simple argument proves that any definite ternary form must fail to represent infinitely many integers, even over the rationals. For if a ternary form f of determinant d represents anything in the p-adic squareclass of $-d$ over $\mathbf{Q}_p$, then it must be p-adically equivalent to $[-d, a, b]$ where the "quotient form" $[a, b]$ has determinant -1, and so p-adically, f must be the isotropic form $[-d, 1, -1]$.

But a positive definite form fails to represent -1, and so is not p-adically isotropic for $p = -1$. By the global relation, there must be another p for which it is not p-adically isotropic, and so it also fails to represent all numbers in the p-adic square-class of $-d$ for this p too!

The Three Squares Theorem illustrates this nicely—the form $[1, 1, 1]$ fails to represent -1 both -1-adically and 2-adically. In the Third Lecture, we showed that The Little Methuselah Form

$$x^2 + 2y^2 + yz + 4z^2$$

failed to represent 31. We now see that since it fails to represent the -1-adic class of its determinant $-31/4$ (i.e., the negative numbers), it must also fail to represent the infinitely many positive integers in the 31-adic squareclass of $-31/4$.

References

Aside from papers (some quite old) that are referred to in the text, we have included references to some textbooks for further reading. First, for more information and many more references see the compendious [CS], which has more to say on almost all of the topics discussed here, as well as the series [CSI-VI].

For the classical theory of binary quadratic forms, one should certainly peruse the landmark [Gau]. For textbooks, see [Bue], [Cas] (advanced but with a fairly down-to-earth viewpoint), [Eich] (for his theorem on the spinor genus), [Fla] (for a redaction of Gauss), [Kit1] (rather abstract), [Jon] (quite elementary), [Mat], [O'M], [Scha], [Ser] (for an excellent elementary treatment of Hasse-Minkowski and Dirichlet's theorem, as well as the p-adic numbers), and [Wat], as well as the references there.

[Aub] L. Aubry, "Solution de quelques questions d'analyse indéterminée", *Sphinx–Œdipe* 7 (1912) 81–84.

[BGM] M. Berger, P. Gauduchon, E. Mazet, *Spectre d'une variété riemannienne*, Lecture Notes in Math. 194, Springer-Verlag, 1971.

[Bli] H. F. Blichfeldt, "The minimum values of positive quadratic forms in six, seven, and eight variables", *Math. Z.* 39 (1935) 1–15.

[Bue] D. A. Buell, *Binary Quadratic Forms: Classical Theory and Modern Computations*, Springer-Verlag, 1989.

[Cas] J. W. S. Cassels, *Rational Quadratic Forms*, Academic Press, 1978.

[CS] J. H. Conway and N. J. A. Sloane, *Sphere Packings, Lattices, and Groups*, Springer-Verlag, 1986 (2nd ed., 1992).

[CS2] J. H. Conway and N. J. A. Sloane, "Four-dimensional lattices with the same theta series", *Duke Math. J.* 66 (1992) 93–96.

[CSI] J. H. Conway and N. J. A. Sloane, "Low-dimensional lattices I: Quadratic forms of small determinant", *Proc. Roy. Soc. Lond. A,* 418 (1988) 17–41.

[CSII] J. H. Conway and N. J. A. Sloane, "Low-dimensional lattices II: Subgroups of $GL_n(\mathbf{Z})$", *Proc. Roy. Soc. Lond. A,* 419 (1988) 29–68.

[CSIII] J. H. Conway and N. J. A. Sloane, "Low-dimensional lattices III: Perfect forms", *Proc. Roy. Soc. Lond. A,* 418 (1988) 43–80.

[CSIV] J. H. Conway and N. J. A. Sloane, "Low-dimensional lattices IV: The mass formula", *Proc. Roy. Soc. Lond. A,* 419 (1988) 259–286.

[CSV] J. H. Conway and N. J. A. Sloane, "Low-dimensional lattices V: Integral coordinates for integral lattices", *Proc. Roy. Soc. Lond. A,* 426 (1989) 211–232.

[CSVI] J. H. Conway and N. J. A. Sloane, "Low-dimensional lattices VI: Voronoi reduction of three-dimensional lattices", *Proc. Roy. Soc. Lond. A,* 436 (1991) 55–68.

[Del] B. N. Delone, "Sur la partition régulière de l'espace à 4 dimensions", *Izv. Akad. Nauk. SSSR Otdel Fiz-Mat.* Nauk 7 (1929) 79-110, 147–164.

[EN] A. Earnest and G. Nipp, "On the theta series of positive definite quaternary quadratic forms", *C. R. Math. Rep. Acad. Sci. Canad.,* 13 No. 1, (1991), pp. 33–38.

[Eich] M. Eichler, *Quadratische Formen und Orthogonale Gruppen,* Grundl. der Math. Wiss. 63, Springer-Verlag, 1952 (2nd ed. 1974).

[Fla] D. Flath, *Introduction to Number Theory,* John Wiley and Sons, 1989.

[GHL] S. Gallot, D. Hulin, J. Lafontaine, *Riemannian Geometry,* Springer-Verlag, 1987. (2nd ed., 1990).

[Gau1] C. F. Gauss, *Disquisitiones Arithmeticae,* Leipzig: Fleischer, 1801. (English trans., Yale Univ. Press, reprinted Springer-Verlag, 1986).

[Gau2] C. F. Gauss, Besprechung des Buchs von L. A. Seeber, "Untersuchung über die Eigenschaften der positiven ternären quadratischen Formen usw.", *Göttingen Gelehrte Anzeigen,* Jul. 9, 1831 = Werke, II, 1876, 188–196.

[Hur] A. Hurwitz, "Über die Reduktion der binären und ternären quadratischen Formen", Math. Ann. 45 (1894) 85-117 = Werke, Bd. II, 157–190.

[Jon] B. W. Jones, *The Arithmetic Theory of Quadratic Forms,* Carus Math. Monographs 10, John Wiley and Sons, 1950.

[Kac] M. Kac, "Can one hear the shape of a drum?", *Am. Math. Monthly,* 73 No. 4 part II, (1966), pp. 1–23.

[Kit1] Y. Kitaoka, *The Arithmetic of Quadratic Forms,* Cambridge Tracts in Math. 106, Cambridge Univ. Press, 1993.

[Kit2] Y. Kitaoka, "Positive definite forms with the same representation numbers", *Arch. Math.,* 28 (1977), pp. 495–497.

[Kne] M. Kneser, "Lineare Relationen zwischen Darstellungsanzahlen quadratischer Formen", *Math. Ann.,* 168, (1967), pp. 31–39.

[KZ] A. Korkine and G. Zolotareff, "Sur les formes quadratiques positives quaternaires", *Math. Ann.* 5 (1872) 581–583.

[KZ] A. Korkine and G. Zolotareff, "Sur les formes quadratiques", *Math. Ann.* 6 (1873) 366–389.

[KZ] A. Korkine and G. Zolotareff, "Sur les formes quadratiques positives", *Math. Ann.* 11 (1877) 242–292.

[Lag] J. L. Lagrange, "Démonstration d'un théorème d'Arithmétique", *Nov. Mém. Acad. Roy. Soc. de Berlin,* ann. 1770 (1772), 123–133= Œuvres v. 3, 189–201.

[Leg] A. M. Legendre, *Essai sur la théorie des nombres,* Paris: Chez Duprat, 1798.

[Mat] G. B. Mathews, *Theory of Numbers,* 1896 (reprinted Chelsea, 1961).

[MH] J. Milnor and D. Husemoller, *Symmetric Bilinear Forms,* Ergebnisse der Math. 73, Springer-Verlag, 1973.

[Mor] L. J. Mordell, "Observation on the minimum of a positive definite definite quadratic forms on eight variables", *J. Lond. Math. Soc.* (1944) 3–6.

[O'M] O'Meara, O. T., *Introduction to Quadratic Forms,* Grundl. der Math. Wiss. 117, Springer-Verlag, 1963 (2nd printing, 1971).

[Scha] W. Scharlau, *Quadratic and Hermitian Forms,* Grundl. der Math. Wiss. 270, Springer-Verlag, 1985.

[Sch1] A. Schiemann, "Ein Beispiel positiv definiter quadratischer Formen der Dimension 4 mit gleichen Darstellungszahlen". *Arch. Math.* 54 (1990), 372–375.

[Sch2] A. Schiemann, "Ternäre positiv definite quadratische Formen mit gleichen Darstellungszahlen", Dissertation, Bonn, 1993.

[Scho] A. Scholz, *Einführung in die Zahlentheorie,* Sammlung Göschen Bd. 1131, W. de Gruyter & Co., 1939.

[Sel] E. Selling, "Über die binären und ternären quadratischen Formen", *J. reine angew. Math.* 77 (1874), 143–229.

[Ser] J. P. Serre, *Cours d'Arithmétique,* Paris: Presses Universitaires de France, 1970 (English trans: *A Course in Arithmetic,* Grad. Texts in Math. 7, Springer-Verlag, 1973).

[Smi1] H. J. S. Smith, "On the orders and genera of ternary quadratic forms", *Phil. Trans. Roy. Soc. Lond.*, 157 (1867) 255-298 = *Coll. Math. Papers,* Vol. I, 455–509.

[Smi2] H. J. S. Smith, "On the orders and genera of quadratic forms containing more than three indeterminates", *Proc. Roy. Soc.* 16 (1867) 197-208 = *Coll. Math. Papers* Vol. I, 510–523.

[Sto] M. I. Stogrin, "Regular Dirichlet-Voronoi partitions for the second triclinic group", *Proc. Stekl. Inst. Math.* #123, 1973.

[ST] I. Stewart and D. Tall, *Algebraic Number Theory,* 2nd. ed., Chapman and Hall, 1987.

[Vet] N. M. Vetčinkin, "Uniqueness of the classes of positive quadratic forms on which the values of Hermite constants are obtained for $6 \leq n \leq 8$", *Proc. Stekl. Inst. Math.* #152 (1982) 37–95.

[Vor] G. F. Voronoï, "Nouvelles applications des paramètres continus à la théorie des formes quadratiques", *J. reine. angew. Math.*: I. Sur quelques propriétés des formes quadratiques positives parfaites, 133 (1908) 97–178; II.1 Recherches sur les paralléloèdres primitifs, 134 (1908) 198–287; II.2 Domaines de formes quadratiques correspondant aux différents types de paralléloèdres primitifs, 136 (1909) 67–181.

[Wat] G. L. Watson, *Integral Quadratic Forms,* Cambridge Tracts in Math. and Math. Phys. 57, Cambridge Univ. Press., 1960.

[Wi] E. Witt, "Eine Identität zwischen Modulformen zweiten Grades", *Abhand. Math. Sem. Hamb.,* 14 (1941), pp. 323–337.

Index

A_n lattice, 54
D_n lattice, 55
E_8 lattice, 55
p-adic Gauss means, 108
    invariance of, 110
p-adic antisquare, 95
p-adic integer, 118
p-adic numbers, 117, 119, 121, 123, 125
p-adic rational equivalence, 97
p-adic rationals, 91, 97, 119
p-adic signatures, 94, 117
    invariance of, 117
p-adic squareclasses, 120
p-adic symbols, 105
    and the genus, 107
    audibility of, 52, 110
    invariance of, 110
p-excesses, 97, 133
p-terms, 100
$\mathbf{GL}_2(\mathbf{Z})$, 31
$\mathbf{PGL}_2(\mathbf{Z})$, 33
$\mathbf{PSL}_2(\mathbf{Z})$, 27, 29, 31, 33
$\mathbf{PSL}_2(\mathbf{Z})$, 27, 37
$\mathbf{SL}_2(\mathbf{Z})$, 31
$\mathbf{Z}$-module, 3
Štogrin, M. I., 85
2-adic Gauss means, 112
2-adic antisquares, 96
2-adic symbol
    invariance of, 112
2-adic symbols, 105, 106
2-signatures, 96
3-dimensional lattices, 69

afterthoughts, 27
Albers, Donald, ix
algorithm, 25
    for equivalence problem, 25
    for representation problem, 25
ambiguous forms, 121

antisquares
    p-adic, 95
    2-adic, 96
Apollonian identity, 99
Apollonius, theorem of, 8
arithmetic progression, 9
arithmetic progression rule, 8, 23
arrows, in topograph, 9
Aubry, L., 138
audibility, 36, 48, 91
    for 2-dimensional lattices, 45
    for 3-dimensional lattices, 45
    of p-adic symbols, 52, 110
    of p-dimensions, 51
    of cubicity, 42
    of determinant, 48
    of genus in 4 dimensions, 114
    of genus in 4 dimensions, 112
    of theta functions, 49

Baranovskii, 89
bases, 5
    Minkowski reduced, 81
binary forms over $\mathbf{Q}_p$, 119
Blichfeldt, H. F., 80
Brillouin zone, 62
Buser, 45
Buser, Peter, 35

characters, 65
    conorms of, 69
    proper, 66
characters and conorms, 65
classification
    of forms, 18
    of indefinite forms, 61
    of integral binary quadratic forms, 26
    of quadratic forms, 53
Climbing Lemma, 11
codes, isospectral, 41

conjugate Minkowski unit, 103
connectedness of topograph, 12, 15
conorms, 15, 61, 65, 69, 76, 85
    and Selling parameters, 76, 85
    characters and, 65
    for 3-dimensional lattices, 69
    for 4-dimensional lattices, 85
    of characters, 69
    putative, 74
    recovering vonorms from, 69
    space of, 66
cubic lattice, 39, 114
cubicity
    audibility of, 42, 60
    inaudibility of, 42
cuboidal box, 79

Davidoff, Guiliana, ix
definite ternary forms
    non-unversality of, 142
Delone, B. N., 76, 85
determinant, 5, 48
    audibility of, 48
determinant of a lattice, 3
diagonal form
    Gauss mean of, 51
diagonalization, 92
diamond packing, 38
Diophantine approximation, 30
Diophantine equation, 22, 25
direct sum
    Gauss mean of, 50
Dirichlet boundary condition, 47
Dirichlet cell, 62
Dirichlet isospectrality, 48
Dirichlet problem, 35
Dirichlet theorem, 121
discriminant, 3
double wells, 15, 17, 65
drum, 35
    shape of, 35, 45
dual lattice, 49
dual quotient group, 49, 54

Eichler, M., 124
    classification theorem, 61, 125
eigenfunction, 46
eigenvalues, 35

Engel, 89
equivalence
    p-adic rational, 97
    rational, 92
    with inessential denominator, 123
equivalence problem, 25
essential prime, 123
even unimodular forms, 136
even unimodular lattices, 38, 53
    24-dimensional, 57
    history of, 136
exclusive Hilbert product, 103

face, 7
Fano plane, 67
Farey fractions, 27, 29, 31, 33
Farey series, 29
Ford circles, 29
forms
    $0+-$, 24
    $0, 0+, 0-, 0+-$, 18
    ambiguous, 121
    indefinite, 18
    integer-valued, 20
    isotropic, 121
    Little Methuselah, 81
    Minkowski reduced, 81
    non-universality of definite ternary, 142
    not representing 0, 18
    representing 0, 24
    semidefinite, 23
    universality of, 141
    with a weir, 25
four-dimensional graphical lattices, 86
fundamental regions, 28
Fung, Francis, ix

Gauss mean, 49
    invariance of p-adic, 110
Gauss means, 50
    p-adic, 108
    2-adic, 112
    of diagonal forms, 51
    redefined, 108
Gauss, C. F., 1, 50, 80, 121, 139
genus, 61, 107
    and p-adic symbols, 107

audibility of, in 4 dimensions, 112, 114

inaudibility of, in higher dimensions, 114

geometry

hyperbolic, 28

Gillman, Leonard, ix

global relation, 97, 117, 122

proof of, 133

glue vectors, 54

gluing lattices, 56

gluing method, 53, 55, 57, 59

Gordon, Caroline, 35, 45

graphical 4-dimensional lattices, 86

Hasse-Minkowski, 102

theorem, proof of, 98

invariants, 102

strong principle of, 135

theorem of, 96

Hedrick, Earle Raymond, vii

Hensel, K. , 117

hexagonal lattice, 37

hexagonal prism, 78

hexarhombic dodecahedron, 78

higher-dimensional, 36

Hilbert norm residue symbol, 103

Hilbert product, versions of, 103

horocyle, 33

hyperbolic geometry, 28

hyperbolic plane, 33

improper vonorms, 65

inaudibility

of cubicity, 42

of genus in higher dimensions, 114

inclusive Hilbert product, 103

indefinite forms, 18, 61

classification of, 61

representing 0, 24

inessential denominator, 123

integer-valued, 3

integer-values form, 20

integral binary forms

classification of, 26

integral equivalence, 4, 112

integral forms

invariants of, 91, 103

with prescribed invariants, 122

integral quadratic form, 1, 3

integrally equivalent, 4

invariance of the p-adic signatures, 117

invariants

for integral forms, 103

trivial, 98

isocubic lattice, 40, 114

isometry group, 22, 25

isospectral

torus, 38

codes, 41

domains, 45

manifolds, 45

isospectral lattices, 36, 38, 40

12- and 8-dimensional examples, 40

16-dimensional, 38

6-dimensional, 40

6-dimensional examples, 41

infinite families of 5-dimensional examples, 42

Milnor's examples of, 37

nonexistence of 2-dimensional examples, 44

nonexistence of 3-dimensional examples, 45

Schiemann's 4-dimensional examples, 42

Schiemann's first pair of, 44

tetralattices, 42

isotropic forms, 121

Jacobi, 39

formula of, 49

Jacobi symbol, 106, 127

linearity of, 132

periodicity of, 131

reciprocity for, 131

Jacobi, C. G., 39

Jordan constituents, 105

Jordan decomposition, 105

Kac, Mark, 35, 45

Kitaoka, Y., 40

Kneser, M., 40, 53, 55, 57, 59

Korkine, A., 80, 137

Lagrange, J.-L., 141

Four Squares Theorem of, 141
lake, 18, 23
 and river, 24
Laplacian, 46
lattice, 2, 36
 A_n, 54
 D_n, 55
 E_8, 55
 dual quotient group of, 49
 gluing, 56
 isocubic, 40
 Leech, 57, 137
 Voronoi cell of, 61
lattices
 16-dimensional, 38
 4-dimensional graphical, 86
 even unimodular, 38, 53
 history of even unimodular, 136
 isospectral, 38, 40
 Niemeier's 24-dimensional, 56, 57
 non-graphical, 89
 primitive, 89
 sphere packing, 80
 vonorms and conorms for 3-dimensional lattices, 69
lax, 5
 base, 5
 superbase, 5
 vector, 5
lax Voronoi vectors, 64
laxly obtuse superbases, 65
Leech lattice, 57
Leech, John, 57, 137
Legendre symbol, 103, 127
Legendre, A. M., 1, 121, 137
Lie algebra of type E_8, 38
linear Jacobi symbols, 132
Little Methuselah Form, 81

manifolds, Riemannian, 35, 36
matrix, 3, 4
 from topograph, 10
matrix-integral, 3
mediant fraction, 29
Milnor, John, 35
minimal norm, 80
Minkowski reduced bases or forms, 81

Minkowski reduction, 80, 83
Minkowski unit, 102
 conjugate, 103
modular form, 37
modular group, 37
Mordell, L. J., 80

negative definite, 13
negative semidefinite, 13
Neumann boundary condition, 48
Niemeier's 24-dimensional lattices, 56
Niemeier, H.-V., 57, 137
norms of vectors, 8

obtuse superbases, 62, 69
 in 3 dimensions, 71
oddity, 96
oddity-fusion rule, 107
orthogonal group, 123

periodic river, 20
periodicity of the Jacobi symbol, 131
positive definite, 13
positive definite forms in high dimensions, 53
positive semidefinite, 13
primitive lattices, 89
primitive vectors, 5
 in the topograph, 7
product formula, 97
proper characters, 66
proper vonorms, 65, 69
putative conorms, 74
putative vonorms, 76

quadratic forms, 1, 18
 classification of, 53
 direct sum of, 50
 Gauss means of, 49
 with respect to a base, 10
quadratic reciprocity, 127
quotient group $L/2L$, 63
quotient manifold, 36

rational equivalence, 92
 complete theory of, 91
rational forms, 91
 with prescribed invariants, 121
rationals, p-adic, 97

reciprocity for the Jacobi symbol, 131
rectangular parallelepiped, 79
Replacement Lemma, 100
rhombic dodecahedron, 78
Riemannian manifolds, 35, 36
river, 18, 19
    between two lakes, 24
    of zero length, 24
    periodic, 20
    uniqueness of, 20
river edges, 19
root lattices, 54, 58
Ryshkov, 89

Schiemann, 42
Schneeberger, William, 141
Scholz, A., 132
Selling parameters, 15, 62, 76, 85
    and conorms, 76, 85
    for 4-dimensional lattices, 85
Selling's formula, 14, 16
Selling, E., 15, 62, 76
semidefinite forms, 23
Serre, J. P., 121
shape, 35
    of a drum, 35, 45
    of a lattice, 35, 37, 39, 41, 43, 45,
        47, 49, 51
sign-walking rule, 107
signatures
    p-adic, 94
    2-adic, 96
    Sylvester's, 95
simple wells, 15, 16, 65
Sloane, Neil, 40
Smith, H. J. S., 137
sphere packing lattices, 80
spinor equivalence, 124
spinor genus, 61, 124
spinor kernel, 124
spinor norm, 124
squareclasses, 119
    p-adic, 120
strict, 5
    base, 5
    superbase, 5
    vector, 5

strict Voronoi vectors, 63
strictly obtuse superbases, 65
Sunada, Toshikazu, 35
superbases, 5
    laxly obtuse, 65
    obtuse, 62
    strictly obtuse, 65
superbases, obtuse, 69
Sylvester, J. J., 95
symbol, 112

tetracode, 42
tetralattices, 42
theta functions, 36, 49
    audibility of, 49
    for 2-dimensional lattices, 45
theta-constants, 39
Three Squares Theorem, 121, 137
    proof of, 138
topograph, 6–8, 27, 82
    arrows in, 9
    connectedness of, 12, 15
    matrix from, 10
    of Little Methuselah form, 82
    shape of, 8
    tree property of, 10
torus, 36
    isospectral, 38
transplantation proof, 48
tree, 10, 21
tree property of topograph, 10
triangle inequality, 14
trivial invariants, 98
truncated octahedron, 77
Type I, 106
Type II, 106

unit forms, 105
universality of forms, 141
universalityof forms
    lack of, for definite ternary forms,
        142
upper half plane, 27

Vetčinkin, N. M., 80
vonorms, 15, 61, 65, 69
    for 3-dimensional lattices, 69
    improper, 65

proper, 65
putative, 76
recovered from conorms, 69
space of, 66
Voronoi cell, 61, 62, 76, 89
    2-dimensional, 63
    and conorms, 76
    five shapes of, 76
    of 2-dimensional lattice, 76
    of 3-dimensional lattice, 77
    of the origin, 62
    shapes for 4- and 5-dimensional lat-
        tices, 83
    with 62 faces, 89
Voronoi norms, 15
Voronoi vectors, 15, 62
    lax, 64
    strict, 63
Voronoi, G. F., 15, 63, 76

Watson's theorem, 125
Watson, G. L., 124
Webb, 45
Webb, David, 35
weir, 24
well lemma, 14
wells, 13
    double, 15, 17
    for positive definite forms, 13
    simple, 15, 16
Witt's cancellation law, 98
Witt's lemma on root lattices, 58
Witt, Ernst, 38, 57
Wolpert, 45
Wolpert, Scott, 35

Zolotareff, G., 80, 127, 137